AF357835

Obstructive Sleep Apnea (OSA)

Obstructive Sleep Apnea (OSA)

Guest Editor

Konstantinos Chaidas

Basel • Beijing • Wuhan • Barcelona • Belgrade • Novi Sad • Cluj • Manchester

Guest Editor
Konstantinos Chaidas
Ear, Nose, and Throat (ENT) Department
Democritus University of Thrace
Alexandroupolis
Greece

Editorial Office
MDPI AG
Grosspeteranlage 5
4052 Basel, Switzerland

This is a reprint of the Special Issue, published open access by the journal *Life* (ISSN 2075-1729), freely accessible at: www.mdpi.com/journal/life/special_issues/O_S_A.

For citation purposes, cite each article independently as indicated on the article page online and using the guide below:

Lastname, A.A.; Lastname, B.B. Article Title. *Journal Name* **Year**, *Volume Number*, Page Range.

ISBN 978-3-7258-2648-3 (Hbk)
ISBN 978-3-7258-2647-6 (PDF)
https://doi.org/10.3390/books978-3-7258-2647-6

Contents

About the Editor

Konstantinos Chaidas

Konstantinos Chaidas has been an Assistant Professor in Otorhinolaryngology at Democritus University of Thrace, School of Medicine in Alexandroupolis, Greece, as well as an Honorary ENT Consultant at the Imperial College Healthcare NHS Trust in London, UK. He has a special interest in obstructive sleep-disordered breathing and has published widely, mainly in this field. He was awarded a Ph.D. in sleep apnea in 2010. He undertook two-year fellowship training in robotic ENT surgery for sleep apnea and thyroid/parathyroid disease at St. Mary's Hospital in London. He is a fellow of the Royal College of Surgeons of England/FRCS (ORL-HNS).

Preface

Obstructive sleep apnea (OSA) is the most common form of sleep-disordered breathing and is characterized by recurrent episodes of complete or partial upper airway obstruction during sleep, resulting in oxygen desaturation, autonomic dysfunction, and sleep fragmentation. It can affect both children and adults, and the main clinical symptoms include loud snoring, noticeable apneas, and breathing difficulties during sleep.

The purpose of this Special Issue entitled "Obstructive Sleep Apnea (OSA)" is to contribute to a better understanding of this underdiagnosed and often underestimated medical condition by bringing to light recent developments.

Konstantinos Chaidas
Guest Editor

MDPI

Editorial

Editorial on the Special Issue "Obstructive Sleep Apnea (OSA)"

Konstantinos Chaidas [1,2]

1   Ear, Nose, and Throat Department, University Hospital of Alexandroupolis, School of Medicine, Democritus University of Thrace, 68100 Alexandroupolis, Greece; konchaidas@gmail.com
2   Ear, Nose, and Throat Department, Imperial College Healthcare NHS Trust, London W2 1NY, UK

Citation: Chaidas, K. Editorial on the Special Issue "Obstructive Sleep Apnea (OSA)". *Life* **2024**, *14*, 669. https://doi.org/10.3390/life14060669

Received: 15 May 2024
Accepted: 21 May 2024
Published: 23 May 2024

Obstructive sleep apnea (OSA) is the most common form of sleep-disordered breathing and is characterized by recurrent episodes of complete or partial upper airway obstruction during sleep, resulting in oxygen desaturation, autonomic dysfunction and sleep fragmentation. It can affect both children and adults, and the main clinical symptoms include loud snoring, noticeable apneas and breathing difficulties during sleep. Overnight polysomnography is the gold-standard method for diagnosing OSA. Although OSA is common, it is a frequently unrecognized cause of serious disabilities that has serious health and social consequences. If untreated, OSA may cause impaired cognitive ability, road traffic accidents, cardiovascular morbidity and all-cause mortality [1].

Various therapeutic options exist. CPAP is the standard treatment for adult OSA, although its clinical application can be compromised by intolerance and poor compliance [2], while adenotonsillectomy is the primary treatment option for children with OSA and adenotonsillar hypertrophy [3].

The purpose of this Special Issue entitled "Obstructive Sleep Apnea (OSA)" is to contribute to a better understanding of this underdiagnosed and often underestimated medical condition by bringing to light recent developments. The collection contains ten articles in the form of seven original studies and three review articles.

Despite recent advances in our understanding of OSA, certain aspects of the disease's pathophysiology have yet to be fully elucidated. The cyclic pattern of intermittent hypoxia in OSA triggers oxidative stress, contributing to cellular damage. The review by Lavalle et al. (Contribution 7) explored the relationship between OSA and oxidative stress, shedding light on the molecular mechanisms involved and some potential therapeutic interventions.

If left untreated, OSA represents a significant mortality risk. Among other morbidities, OSA is considered to be a risk factor for erectile dysfunction. The study by Martynowicz et al. (Contribution 8) found that subjects with erectile dysfunction have altered sleep architecture, oxygen saturation parameters and increased daytime sleepiness.

An early and accurate diagnosis of OSA is of paramount importance and, although overnight polysomnography remains the gold-standard diagnostic tool, a detailed patient history is always valuable. Several questionnaires have been developed to assist in the process of screening patients with suspected OSA and one of the simplest and most frequently used is the Epworth Sleepiness Scale (ESS), which measures daytime somnolence. This parameter is often measured differently by patients and their partners and there is still confusion regarding the utility of partner-completed ESS in identifying OSA and predicting its severity. Chaidas et al. (Contribution 3) showed that there is a strong correlation between patient- and partner-reported ESS scores, but neither patient- nor partner-completed ESS were associated with OSA severity.

A thorough clinical examination of the upper airway is essential in patients with OSA as the findings can help in guiding appropriate treatment. This Special Issue presents some new suggestions regarding the assessment of patients with OSA. Specifically, Morato et al. (Contribution 1) suggested a new tool for palatopharyngeal muscle assessment during intraoral examination, which may be useful for creating a common language for sleep surgeons. Moreover, the use of artificial intelligence (AI) in medicine is becoming increasingly popular. A systematic review by Tsolakis et al. (Contribution 9) demonstrated that automatic airway segmentation is accurate, fast and easy to use in the measurement of airways. Thus, the future use of AI in assessing airway patency in OSA patients is promising at the very least.

CPAP is the first-line treatment for adult OSA but its efficacy may be limited by the variability that is seen in the rates of compliance with such treatment. For that reason, precise treatment assessment is particularly important. Brajer-Luftmann et al. (Contribution 4) showed that the automatic algorithm that is used in auto-CPAP measurement is a good tool for the assessment of the treatment efficacy of CPAP in home settings.

In addition to CPAP, various alternative therapeutic options for OSA exist in published guidelines worldwide, including lifestyle changes, oral appliances and surgery, with a remarkable variation in their availability across different countries [4,5].

Weight loss plays an important role in OSA management with the aim of at least improving its severity. Considering the fact that OSA and systemic inflammation typically coexist within a vicious cycle, the study by Georgoulis et al. (Contribution 5) explored the effectiveness of a weight-loss lifestyle intervention in reducing plasma tumor necrosis factor-alpha (TNF-a), a well-established modulator of systematic inflammation, and concluded that a healthy lifestyle intervention may reduce systemic inflammation in patients with OSA.

Furthermore, OSA is often associated with craniofacial and orthodontic abnormalities, especially in children, and may require a maxillofacial and/or other orthodontic intervention in selective cases. Caruso et al. (Contribution 2) evaluated the outcomes of orthodontic treatment with a rapid maxillary expander in association with a Delaire mask, demonstrating improvements in airway patency and OSA-related clinical conditions in children with a class III malocclusion.

Tonsillectomy and uvulopalatopharyngoplasty (UPPP) are common procedures that are used in the surgical management of OSA. A study by Hu et al. (Contribution 6) showed that combining a tonsillectomy and UPPP in patients with OSA did not increase the risk of patients developing a deep neck infection in the long term, but may reduce its severity by decreasing the intubation rate and the length of hospitalization.

Although adenotonsillar hypertrophy is the primary cause of OSA in children, the airway pathology in adults is usually more complex with the presence of a multi-level obstruction. The critical role of epiglottis in airway narrowing has recently been revealed. A systematic review by Vallianou and Chaidas (Contribution 10) evaluated surgical treatment options for epiglottic collapse, demonstrating that all of the currently available surgical techniques are safe and effective in managing selected patients. Effective management of epiglottic collapse can improve OSA severity or even cure OSA, but can also improve CPAP compliance. The selection of an appropriate surgical technique should be part of an individualized, patient-specific therapeutic approach.

In summary, this Special Issue offers further insights into the pathophysiology, diagnostic assessment and management of patients with OSA, and highlights the importance of continuous research in this field.

Conflicts of Interest: The author declares no conflicts of interest.

List of Contributions

1. Morato, M.; Cardona-Sosa, M.P.; Bosco, G.; Pérez-Martín, N.; Marte-Bonilla, M.M.; Marco, A.; O'Connor-Reina, C.; Lugo, R.; Plaza, G. Palatopharyngeal Arch Staging System (PASS): Consensus about Oropharyngeal Evaluation. *Life* **2023**, *13*, 709. https://doi.org/10.3390/life13030709.
2. Caruso, S.; Lisciotto, E.; Caruso, S.; Marino, A.; Fiasca, F.; Buttarazzi, M.; Sarzi, A.m.a.d.è.; D; Evangelisti, M.; Mattei, A.; Gatto, R.; et al. Effects of Rapid Maxillary Expander and Delaire Mask Treatment on Airway Sagittal Dimensions in Pediatric Patients Affected by Class III Malocclusion and Obstructive Sleep Apnea Syndrome. *Life* **2023**, *13*, 673. https://doi.org/10.3390/life13030673.

3. Chaidas, K.; Lamprou, K.; Stradling, J.R.; Nickol, A.H. Association between Patient- and Partner-Reported Sleepiness Using the Epworth Sleepiness Scale in Patients with Obstructive Sleep Apnoea. *Life* **2022**, *12*, 1523. https://doi.org/10.3390/life12101523.

4. Brajer-Luftmann, B.; Trafas, T.; Mardas, M.; Stelmach-Mardas, M.; Batura-Gabryel, H.; Piorunek, T. The Automatic Algorithm of the Auto-CPAP Device as a Tool for the Assessment of the Treatment Efficacy of CPAP in Patients with Moderate and Severe Obstructive Sleep Apnea Syndrome. *Life* **2022**, *12*, 1357. https://doi.org/10.3390/life12091357.

5. Georgoulis, M.; Yiannakouris, N.; Tenta, R.; Kechribari, I.; Lamprou, K.; Vagiakis, E.; Kontogianni, M.D. Improvements in Plasma Tumor Necrosis Factor-Alpha Levels after a Weight-Loss Lifestyle Intervention in Patients with Obstructive Sleep Apnea. *Life* **2022**, *12*, 1252. https://doi.org/10.3390/life12081252.

6. Hu, P.C.; Shih, L.C.; Chang, W.D.; Lai, J.N.; Liao, P.S.; Tai, C.J.; Lin, C.D.; Yip, H.T.; Shen, T.C.; Tsou, Y.A.; et al. The Changes in the Severity of Deep Neck Infection Post-UPPP and Tonsillectomy in Patients with OSAS. *Life* **2022**, *12*, 1196. https://doi.org/10.3390/life12081196.

7. Lavalle, S.; Masiello, E.; Iannella, G.; Magliulo, G.; Pace, A.; Lechien, J.R.; Calvo-Henriquez, C.; Cocuzza, S.; Parisi, F.M.; Favier, V.; et al. Unraveling the Complexities of Oxidative Stress and Inflammation Biomarkers in Obstructive Sleep Apnea Syndrome: A Comprehensive Review. *Life* **2024**, *14*, 425. https://doi.org/10.3390/life14040425.

8. Martynowicz, H.; Poreba, R.; Wieczorek, T.; Domagala, Z.; Skomro, R.; Wojakowska, A.; Winiewska, S.; Macek, P.; Mazur, G.; Gac, P.; et al. Sleep Architecture and Daytime Sleepiness in Patients with Erectile Dysfunction. *Life* **2023**, *13*, 1541. https://doi.org/10.3390/life13071541.

9. Tsolakis, I.A.; Kolokitha, O.E.; Papadopoulou, E.; Tsolakis, A.I.; Kilipiris, E.G.; Palomo, J.M. Artificial Intelligence as an Aid in CBCT Airway Analysis: A Systematic Review. *Life* **2022**, *12*, 1894. https://doi.org/10.3390/life12111894.

10. Vallianou, K.; Chaidas, K. Surgical Treatment Options for Epiglottic Collapse in Adult Obstructive Sleep Apnoea: A Systematic Review. *Life* **2022**, *12*, 1845. https://doi.org/10.3390/life12111845.

References

1. Marshall, N.S.; Wong, K.K.; Liu, P.Y.; Cullen, S.R.; Knuiman, M.W.; Grunstein, R.R. Sleep apnea as an independent risk factor for all-cause mortality: The Busselton Health Study. *Sleep* **2008**, *31*, 1079–1085. [PubMed]

2. Brimioulle, M.; Chaidas, K. Nasal function and CPAP use in patients with obstructive sleep apnoea: A systematic review. *Sleep Breath.* **2022**, *26*, 1321–1332. [CrossRef] [PubMed]

3. Marcus, C.L.; Moore, R.H.; Rosen, C.L.; Giordani, B.; Garetz, S.L.; Taylor, H.G.; Mitchell, R.B.; Amin, R.; Katz, E.S.; Arens, R.; et al. A randomized trial of adenotonsillectomy for childhood sleep apnea. *N. Engl. J. Med.* **2013**, *368*, 2366–2376. [CrossRef] [PubMed]

4. Mandavia, R.; Mehta, N.; Veer, V. Guidelines on the surgical management of sleep disorders: A systematic review. *Laryngoscope* **2020**, *130*, 1070–1084. [CrossRef] [PubMed]

5. Chaidas, K.; Ashman, A. Variations in funding for treatment of obstructive sleep apnoea in England. *J. Laryngol. Otol.* **2021**, *135*, 385–390. [CrossRef]

Article

Palatopharyngeal Arch Staging System (PASS): Consensus about Oropharyngeal Evaluation

Marta Morato [1,2], Maribel P. Cardona-Sosa [3,4], Gabriela Bosco [1,5], Nuria Pérez-Martín [1,5], Mayerin M. Marte-Bonilla [3,4], Alfonso Marco [6], Carlos O'Connor-Reina [7], Rodolfo Lugo [3,4] and Guillermo Plaza [1,5,*]

1   Department of Otolaryngology, Hospital Universitario de Fuenlabrada, Universidad Rey Juan Carlos, 28942 Madrid, Spain
2   Department of Otolaryngology, Hospital Quirónsalud Madrid, 28223 Madrid, Spain
3   Ronquido Monterrey, Centro de Diagnóstico y Tratamiento, Monterrey 64660, Mexico
4   Department of Otolaryngology, Clínica Hospital Constitución, ISSSTE, Monterrey 64530, Mexico
5   Department of Otolaryngology, Hospital Universitario Sanitas La Zarzuela, 28942 Madrid, Spain
6   Department of Otolaryngology, Hospital Universitario Reina Sofía, 30003 Murcia, Spain
7   Department of Otolaryngology, Hospital Quirónsalud Marbella, 29603 Marbella, Spain
*   Correspondence: gplaza.hflr@salud.madrid.org

Abstract: Intraoral examinations are essential in the evaluation of the upper airway in patients with obstructive sleep apnea (OSA). The morphology of the anatomic structures of the soft palate, the tonsillar fossae, and the palatoglossus and palatopharyngeal muscles is an important determinant of the size and collapsibility of the velum and oropharynx. The Palatopharyngeal Arch Staging System (PASS) is a systematic way to explore the oropharynx and report anatomic variations in the visible part of the palatopharyngeal muscle. In this prospective study, 30 sleep surgeons evaluated the reliability of the PASS using a selection of 23 videos of oropharyngeal examinations of healthy patients. The corresponding score on the PASS scale was graded for each examination. For internal structure and internal agreement, the Cronbach and Krippendorff alpha values were 0.96 and 0.46, which corresponded to a nearly perfect interrelationship and a moderate agreement, respectively. These findings suggest that the PASS is a valuable tool for evaluating the position of the palatopharyngeus muscle during oropharyngeal examinations and may be useful for creating a common language for sleep surgeons when evaluating the palatopharyngeal muscle.

Keywords: interexaminer agreement; obstructive sleep apnea; palatopharyngeal arch; staging system

Citation: Morato, M.; Cardona-Sosa, M.P.; Bosco, G.; Pérez-Martín, N.; Marte-Bonilla, M.M.; Marco, A.; O'Connor-Reina, C.; Lugo, R.; Plaza, G. Palatopharyngeal Arch Staging System (PASS): Consensus about Oropharyngeal Evaluation. *Life* **2023**, *13*, 709. https://doi.org/10.3390/life13030709

Academic Editor: Konstantinos Chaidas

Received: 29 December 2022
Revised: 20 February 2023
Accepted: 1 March 2023
Published: 6 March 2023

1. Introduction

Obstructive sleep apnea (OSA) is a highly prevalent disease with a complex pathophysiology that entails several anatomic and functional mechanisms [1]. The morphology of the anatomic structures of the soft palate, the tonsillar fossae, and the palatoglossus and palatopharyngeal muscles is an important determinant of the size and collapsibility of the velum and oropharynx. The structure and obstructive pattern of the pharynx differ between people, and the palatopharynx's muscle position, angulation, and length determine the size and shape of the closure patterns of the retropalatal airway [2]. Therefore, treatment success depends to a large extent on patient selection, which is important for patients undergoing upper airway (UA) surgery [3,4].

The approach to treating OSA is steadily moving from a continuous positive airway pressure-centered "one-size-fits-all" approach to individualized multimodality treatments of UA obstruction [5]. Several staging systems have been proposed for use when planning palatopharyngeal surgeries. Friedman and colleagues correlated oropharyngeal evaluations performed in an office with the prognosis after surgical treatments [3,4]. The current evolution of OSA surgery has moved from simply decreasing the obstruction or enlarging the UA size to stiffening the palate and lateral walls of the oropharynx to minimize the dynamic collapse [5].

The diagnostic approaches and training programs can differ between otolaryngology units and sleep surgeons. Cammaroto et al. [6] reported significant concordance and few divergences of interest in the diagnosis and treatment of sleep disorders between nationalities and types of institutions. Specific scales may be needed to facilitate clear communication between sleep surgeons throughout the world.

Anatomically, the size of the supratonsillar fat pads and the size of the palatine tonsils and their base of implantation in relation to the palatopharyngeus and palatoglossus muscles contribute to individual differences. During presurgical evaluation, it may be helpful to characterize the palatopharyngeus muscle (PPM) because this muscle is involved in most pharyngoplasties. Currently, there is no standardization for its classification as part of physical examinations [7,8].

Morphological Description of Palatopharyngeus Muscle

The PPM is composed of the aponeurosis and the inferior pterygoid bundles, which originate from a region continuing from the palatine aponeurosis to the inferior margin of the medial pterygoid plate. In the palatine aponeurosis, the palatopharyngeal sphincter, which originates from the nasal aspect of the lateral half of the aponeurosis, passes dorsally below the levator veli palatini (LVP). Therefore, it is distinguishable from the nasal fasciculus, which originates from the medial half of the palatine aponeurosis.

This muscle has the following two major divisions: the longitudinal fascicle and the transverse fascicle. The longitudinal fascicle splits into ventral (small) and dorsal (long) parts, which surround the elevator of the palate velum, before combining into a more concise fascicle that involves the transverse fascicle in the palatopharyngeal arch and the lateral wall of the pharynx. The dorsal and ventral longitudinal heads lie medial to the superior constrictor muscle of the pharynx. Okuda et al. [9] reported that the PPM fascicles comprising the palatopharyngeal arch split into the following two directions: nasal (superficial or luminal layer) and oral (deep layer). The PPM has variations in its form, and these contribute to different anatomic phenotypes, as reviewed by Olszewska and Woodson [2].

These merge from the origin into the posterior part of the palatine aponeurosis and extend in an inferior direction up to its insertion into a large area of the pharyngeal wall (Figure 1).

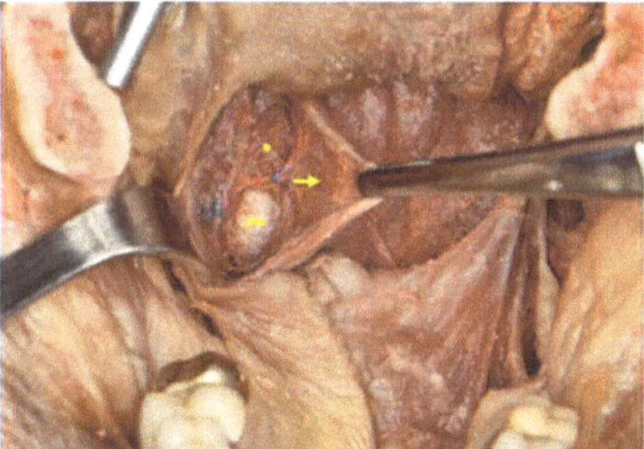

Figure 1. Exposure of the palatopharyngeal muscle (arrow) after medial traction of the posterior pillar and its relationship with the superior constrictor muscle (*) and fascia (**).

The most important functions of the PPM include reducing the size of the pharyngeal isthmus, lowering the palate, and raising the larynx. The transition between the PPM and the superior constrictor of the pharynx is known as the palatopharyngeal sphincter.

There are three movements of the walls of the pharyngeal isthmus concerned with the velopharyngeal closure, which are as follows: the backward movement of the anterior wall, the medial movement of the lateral wall, and the forward movement of the posterior wall. Considering the arrangement of muscles around the pharyngeal isthmus (Figure 1), the elevation of the velum by the LVP is responsible for the movement of the anterior wall, while the palatopharyngeal sphincter (PPS) and the superior constrictor of the pharynx (SCP) are responsible for the movement of the lateral and posterior walls. However, this may be ineffective for the movement of the lateral wall owing to its bilateral attachments being fixed in an immovable position by the pterygoid hamulus.

The contraction of the PPM increases the efficiency of the velopharyngeal closure by exerting pressure on the salpingopharyngeal fold and uvula muscles toward the velum. During swallowing, the function of the PPM is constriction through medialization of the lateral wall and shortening of the pharynx. The transverse fascicle flows dorsally from the soft palate to reach the pharyngeal raphe, forming a ridge known as the Passavant's ridge, through which the soft palate rises to separate the nasopharynx from the oropharynx. Therefore, the PPM acts to raise the pharynx or depress the soft palate and the nasopharyngeal sphincter [8].

However, differences in the position of this muscle in the oropharynx can cause variations in the distance between the contralateral muscle and muscular tone, and these variations contribute to individual differences in anatomic characteristics or functions [10]. Thus, the shape taken by the PPM in its trajectory toward its insertion may help in staging during the preoperative examination of the posterior pillar. The morphology of the fossae palatina and its muscles condition the narrowing of the SPC and contribute to snoring. Lugo et al. [11] designed the Palatopharyngeal Arch Staging System (PASS) as a system for reporting anatomic variations in the PPM during oropharyngeal examinations of patients with OSA. The PASS may be useful for determining the best type of pharyngoplasty for each patient.

The aims of this study were to assess the usefulness of the PASS as a system for reporting anatomic variations in the PPM during oropharyngeal examination and to determine its reliability for standardizing a common language among sleep surgeons.

2. Materials and Methods

A prospective, non-randomized, observational, and longitudinal study was designed. Institutional review board approval was obtained from Hospital Universitario de Fuenlabrada, Madrid, Spain (IRB): 22/116.

2.1. Population

The participants were healthy patients with adequate sleep hygiene and no complaints of snoring or daytime sleepiness (Epworth Sleepiness Scale < 7 points). We choose to describe and evaluate the distribution of the PPM into the oropharynx only in people without OSA. People with tonsil stages 3 or 4 on the Brodsky scale were excluded from this study because the amount of tonsillar tissue may complicate the PPM examination. Those with Friedman tongue position (FTP) grades 3 or 4 were also excluded because these positions may complicate the morphological assessment of the uvula and PPM.

2.2. Staging System

The PASS [11] comprises a five-item scale that reports on the different possible PPM dispositions during oropharyngeal staging on a scale from 0 to 4 (Figure 2). Videos of each PASS type are included as Supplementary Materials.

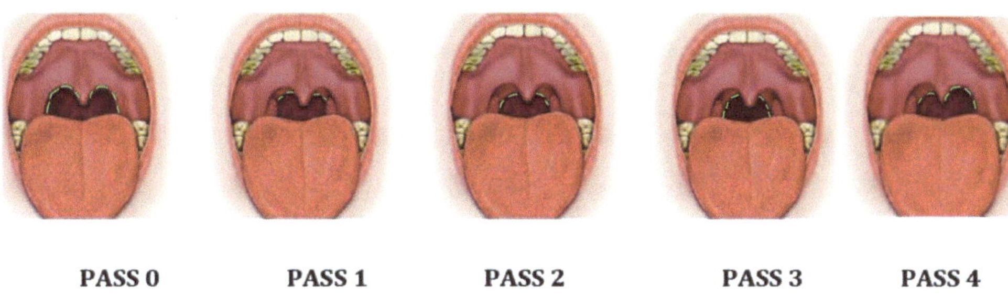

PASS 0   **PASS 1**   **PASS 2**   **PASS 3**   **PASS 4**

Figure 2. Graphic picture of the PASS. See description in the text.

The description of each stage is as follows:

- PASS 0: There has been a surgical procedure involving the oropharynx, such as a tonsillectomy or any type of pharyngoplasty. It is not possible to visualize any posterior pillar (PPM).
- PASS 1: The position of the PPM arises from the uvula to the upper pole of both tonsils, and only the upper part of the PPM muscle can be seen. In this stage, most of the PPM is located behind the tonsillar tissue, probably because of a thin PPM.
- PASS 2: The position of the PPM is observed in the upper two-thirds portion from the uvula to the middle pole of the tonsils. A thicker muscle is observed behind the tonsillar tissue, and the upper half of the PPM can be evaluated intraorally.
- PASS 3: The entire portion of the PPM can be observed through the mouth. A thick and powerful muscle is visible behind the tonsillar tissue.
- PASS 4: Asymmetry between the right and left sides is observed when visualizing the PPM.

2.3. Digital Video Evaluation

Using a brief description and a graphical representation, the PASS was explained to 30 sleep surgeons from different countries, including Chile, Colombia, the Dominican Republic, Ecuador, Israel, Mexico, Perú, and Spain. This number of reviewers was calculated to provide enough observers to have statistical power in the analysis. These reviewers did not have information about the patients' clinical history and included members of the network of specialist sleep surgeons from the Ibero-American Society of Sleep Surgery. A total of 23 videos were attached in digital format (five videos are included as Supplementary Materials as Videos S1–S5), each lasting about 15 s, and obtained during office visits by patients at the Hospital Constitución ISSSTE in Monterrey, Nuevo León, Mexico, in May 2021. These videos were sent to the reviewers, who were allowed free access and unlimited time for repeated viewing. This was intended to correspond to different oropharyngeal examinations performed with specific instructions to assess patients with the support of a lingual depressor and the patient saying "A". The classification was performed using an attached evaluation sheet to determine the scale corresponding to each video. The demographic data, including sex, age, and Brodsky tonsil grade, were collated.

2.4. Statistical Analysis

The data were processed using IBM SPSS Statistics (version 26.0; IBM Corp., Armonk, NY, USA). The categorical variables are presented as totals and percentages, and the quantitative variables are presented as means and standard deviations. The intrarater test–retest and interrater reliability were analyzed and used to compare the PASS. The intrarater test–retest reliability covers two related but different concepts: reliability and agreement. Reliability is the ability of a measure applied twice to the same respondents to produce the same ranking on both occasions. Agreement requires the measurement tool to produce the same exact values twice.

An internal structure analysis of the PASS scale was used with the Cronbach α, except for studying correlations between the response patterns, item difficulty, and α for each. A concordance analysis was also performed using the Krippendorff and factorial analyses, which are tests to measure the extent of agreement between raters using multiple categories for classifying the same group of patients. This method can be applied to assess the reliability and reproducibility of diagnostic categorizations of patients and measure the extent of agreement that occurs beyond what would occur by chance alone.

The data factorial analysis was also used to report the main variance components. The results between 0.7 and 0.9 in the validation tests were deemed to be values of interest. For the factorial analysis, *eigen* components were filtered with a value of <0.45, and the colinear results were ruled out.

3. Results

The subjects were mainly young adults, with a mean age of 39.7 years (a standard deviation of 9.47), and about 52% (12) of them were women. The tonsils were evaluated as follows: grade 0 was 21.7% (5), grade 1 was 60.8% (14), and grade 2 was 17.5% (4) on the Brodsky scale. Every video was evaluated by 30 sleep surgeons, who used the PASS to score each video. Their results are summarized in Table 1 and Figure 3.

There are five cases in which the SD is more than 1 and different from the other cases. When focusing on these cases, we can see that the confusion factor is the validation of PASS 4, because it induces a wide range of variability when any of the pillars is asymmetric. We can explain these differences in the classification with the pictures in Figure 2, with better accuracy in the rest of the items but not with PASS 4, explaining most of the variability.

Table 1. Total PASS score of every video according to the 30 sleep surgeons. SD: standard deviation. * Cases with SD more than 1.

CASES	PASS 0	PASS 1	PASS 2	PASS 3	PASS 4	Median Rate of Score	SD
1	21	0	1	7	1	0.9	1.38 *
2	0	1	4	24	1	2.83	0.51
3	13	1	2	0	14	2.03	1.87 *
4	1	0	2	27	0	2.83	0.57
5	0	14	10	6	0	1.73	0.76
6	3	0	1	26	0	2.67	0.89
7	0	5	21	4	0	1.97	0.54
8	27	3	0	0	0	0.1	0.3
9	0	17	13	0	0	1.43	0.49
10	0	1	13	16	0	2.5	0.55
11	0	0	4	16	10	3.2	0.64
12	2	27	0	0	1	1.03	0.59
13	0	0	17	11	2	2.5	0.61
14	0	21	2	0	7	1.77	1.24 *
15	0	0	1	26	3	3.07	0.35
16	0	0	10	17	2	2.7	0.58
17	2	28	0	0	0	0.93	0.25
18	0	24	6	0	0	1.2	0.39
19	0	1	24	4	1	2.17	0.51
20	13	3	0	1	13	1.93	1.87 *
21	0	15	0	0	15	2.5	1.48 *
22	2	0	7	21	0	2.57	0.79
23	23	5	2	0	0	0.3	0.58

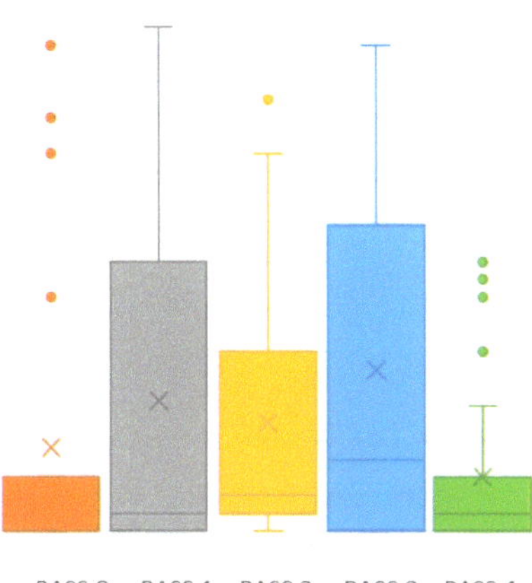

Figure 3. Box-plot distribution of the scores.

The scale's internal structure was evaluated using the Cronbach α, which showed a high level of internal structure design ($\alpha = 0.96$) (Table 2). The Krippendorff test was used to evaluate the extent of the agreement, which was 0.46; in a multiple-choice staging system with more than three options, this range of Krippendorff shows a good correlation. As this was a scale, the Kaiser–Meyer–Olkin test for factorial analysis was used to analyze the response pattern (Figure 4). Twenty-eight records were grouped in the first component of variance. Two records were grouped in the second component of variance, which explained only 15% of the variance and showed fewer interrelationships. The two records that were not included in the first component were detected in this second component. The components of variance 3 (11.04%), 4 (6.78%), and 5 (3.89%) together accounted for a very low variance.

Table 2. Interobserver correlation from each examiner.

Examiner	Interobserver Correlation (IC)	Mean of Response	Alpha Cronbach
1	0.397	2000	0.963
2	0.352	2130	0.963
3	0.787	1826	0.960
4	0.564	2043	0.962
5	0.770	2174	0.960
6	0.551	1957	0.962
7	0.679	2087	0.961
8	0.542	1696	0.962
9	0.838	1957	0.960
10	0.682	2174	0.961
11	0.723	1696	0.961
12	0.685	2087	0.961
13	0.764	1739	0.960
14	0.778	1913	0.960
15	0.495	1565	0.963
16	0.735	2000	0.961
17	0.605	1913	0.962
18	0.488	1957	0.962

Table 2. *Cont.*

Examiner	Interobserver Correlation (IC)	Mean of Response	Alpha Cronbach
19	0.588	1913	0.962
20	0.785	2087	0.960
21	0.550	2261	0.962
22	0.770	1696	0.961
23	0.731	2261	0.961
24	0.641	1826	0.961
25	0.750	2000	0.961
26	0.768	2174	0.960
27	0.707	1696	0.961
28	0.820	2043	0.960
29	0.628	1609	0.962
30	0.769	2043	0.960

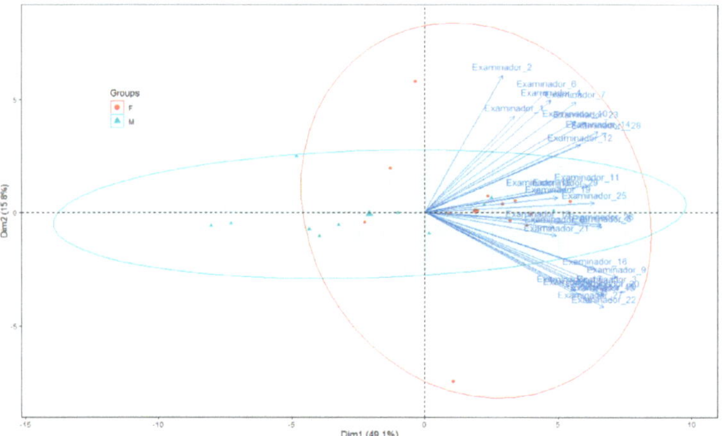

Figure 4. Factorial analysis: distribution of classifications.

The factorial analysis suggested (Figure 4) a high degree of similarity in the response pattern when the 23 records were included in the first component, which accounted for the highest amount of variance in the survey. By the disposition of the response pattern, the two records located in the second component indicated a lack of understanding by two reviewers about how to complete the PASS. This was not detectable in the comparison tests but was found by an analysis of variance.

4. Discussion

The PPM is a key anatomic feature in most pharyngoplasties treating the lateral pharyngeal wall [4]. Therefore, a clear description of the PPM is important for sleep surgeons because the morphology, position of the palatal arches, and muscle tone may influence the pattern of palatal obstruction. The aims of this study were to describe the PASS as a system for reporting anatomic variations in the PPM during oropharyngeal examinations and determine its reliability for standardizing a common language among sleep surgeons. In this manuscript, we have shown its use in healthy subjects.

So far, the most common scale used to evaluate the oropharynx is the Friedman tongue position [3,4]. Friedman et al. [4] performed a prospective study of 172 patients who were being evaluated for OSA. They performed a further modified version of the Mallampati examination, where they asked the patients to sit upright with their heads in a neutral position and open their mouths without sticking their tongues out. They initially called this a "modified Mallampati" grade but later changed the term to "Friedman tongue position (FTP)". They found a statistically significant correlation between the FTP grade and the apnea–hypopnea index (AHI) severity (r = 0.340; $p < 001$). From a clinical perspective, the classification proposed by Friedman et al. is carried out at rest, with the mouth open and the tongue inside the oral cavity. It assesses the tongue distribution with respect to the palate, as well as the tonsil size and BMI. A low Friedman score has been directly related to the possibility of successful surgery following uvulopalatopharyngoplasty (UPPP). However, it does not evaluate the lateral pharyngeal wall.

The main objective of this study was to evaluate the use of the PASS combined with Friedman's classification to help in the selection of a surgical technique for managing the lateral pharyngeal wall. The combination of the Friedman classification with the PASS is one option because the former evaluates the soft palate and its relationship with the tongue and the latter evaluates the lateral pharyngeal wall, PPM, and tonsil size.

Lateral pharyngeal wall surgery can involve several variations of pharyngoplasty [8,11]. For example, Cahali considered hypertrophy of the PPM as a surgical indication for use of the lateral pharyngoplasty technique [12]. Pang and Woodson [13] based their expansion sphincter pharyngoplasty on the need to modify the PPM. Korhan et al. [14] were among the first to examine the relationship between the oropharyngeal anatomy of the posterior palatal arch at the first examination with the degree of snoring and OSA. They concluded that the distance between the PPM was shorter in patients with severe disease. However, they did not comment on the relationship between these findings and the type of pharyngoplasty.

In 2019, Lugo et al. [11] presented the PASS for evaluating anatomic variations in the PPM during oropharyngeal examinations of patients with OSA. Describing the anatomic variations of the palatopharyngeal arch (PASS 0–4) allows for the evaluation of both the interpalatopharyngeal distance and the PPM tone, which may be helpful when deciding on a specific surgical technique for the lateral pharyngeal wall. For example, for patients with a PASS of 0 or 1, lateral pharyngeal wall mobilization will not be the most important issue, whereas for patients with a PASS of 3 or 4, a specific lateral wall pharyngoplasty would be more appropriate for resolving the collapse.

We believe that it is important to be familiar with the morphology and anatomy of the PPM before deciding on the type of surgery for treating OSA. This is because the management of the oropharynx and its lateral pharyngeal wall is a commonly discussed issue, especially as it relates to the type of collapse. An oropharyngeal office examination may allow clinicians to correlate the features of the PPM with PPM intraoperative findings, with the aim of predicting surgical success or guiding the choice of pharyngoplasty. The broad variety of pharyngoplasty and the differences in terms of PPM mobilization are important when dealing with sleep surgery. Specific lateral pharyngeal wall pharyngoplasty procedures can be performed differently, although all methods share the same goal of lateralizing and stiffening the lateral pharyngeal walls [15].

The PASS classification aims to unify a common terminology among sleep surgeons in terms of intraoral classification of findings in the pharyngeal lateral wall. The present study focused on validating the scale and evaluating the reproducibility of its applications in the office by professionals from different countries and backgrounds. Our intention here was to study the extent of interexaminer agreement, similar to the evaluation of the FTP system published by the Friedman group and other authors [16–20].

Our evaluation shows that the PASS has a consistent internal structure and is well designed. The Cronbach α values were high and similar to those of other classifications, such as the Brodsky scale for tonsil size (k = 0.75) and FTP (k = 0.82) [21,22]. However, our Krippendorff α value was low (0.46). These values show that the concordance between evaluators was not ideal and reflect the heterogeneity among sleep surgeons who did not have complete clinical data to make the evaluation.

Each specialist involved in our study evaluated the cases according to his/her own internal criteria, following the guidelines in the brief explanation given with the PASS scale and the graphic outlines given as a visual support. The surgical training for the treatment of OSA may have differed between these specialists given their professional activities in different parts of the world.

The Krippendorff coefficient found here indicated a lower level of agreement than that reported by other papers on the validation of scales, such as that used to measure lymphatic tissue hypertrophy [16]. Despite this, the Cronbach coefficient in our study indicated a very high internal consistency or reliability (0.96).

The reproducibility of staging systems is an important technical requirement for the scale to have an impact. In addition to the clinical feature it reflects, examining the palatopharyngeal muscle is important in terms of OSA [17].

Although the size of the palatine tonsils, as graded using the Brodsky scale [22], can hinder viewing the posterior pillar during oropharyngeal examinations, the exclusion of patients with hypertrophic tonsils grades 3 or 4 should not affect the validity of the classification. According to the recent Spanish OSA consensus [23], the most appropriate treatment for these patients is a tonsillectomy. For this reason, it is not crucial to classify the type of palatopharyngeal muscle during the physical examination of these patients.

As previously described by Friedman, the relationships between OSAS severity according to the Mallampati score, body mass index, size of the tonsils, and the description of and variations in the PPM are each independently related to the presence and severity of OSAS [24,25]. Given that the PPM is an essential muscle for the surgical treatment of OSA, the proposed PASS may be the first step in identifying the risk or likely outcome of palatine velum surgeries (Figure 5 and Video S6 in Supplementary Materials).

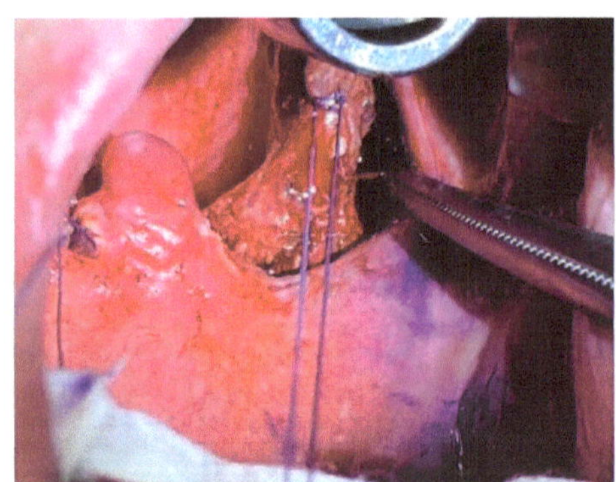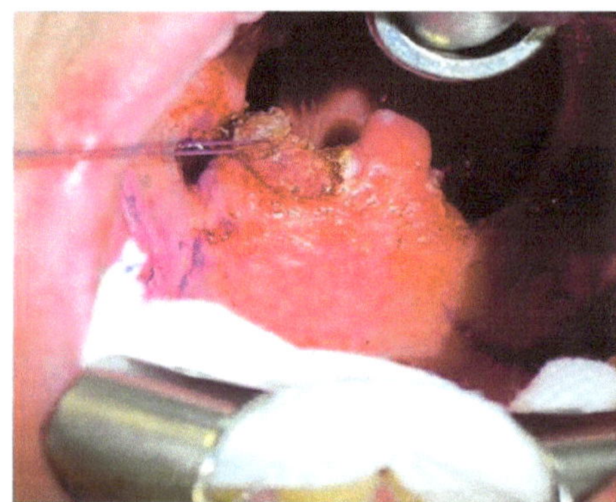

Figure 5. Intraoperative photographs of the palatopharyngeal muscle. In this case, there is a thick palatopharyngeal muscle in both images.

This is even more important when PPM dissection is designed to be performed, either on its own or in combination with some sectioning of the muscle. As some authors are currently promoting, sectioning of the muscle may be indicated in cases with robust PPM, while sparing its anatomy may be an option in cases with thin muscles [7,26]. Thus, studying the PPM prior to or during surgery may be very important to predicting good results.

Our study has some limitations. First, the use of a healthy population allowed for the description of every anatomical possibility without the influence of OSA or any other concurrent phenotype, implying that the evaluation of the PPM was probably modified by other independent factors, such as tongue volume and size, obesity, or nasal obstruction, producing AOS. Our intention is to evaluate this muscle as a self-reliant finding in the oropharynx. Second, the sleep surgeons evaluated the videos but did not examine the patients. This may not have affected the results, but, in some cases, a comprehensive examination of the patient may have been needed. Additionally, as seen in some variations in the interpretation of the videos from the different observers, their different geographical origins may have influenced the results. Third, another limitation is that the external oropharyngeal characteristics of the PPM may be related to the size of the muscle or the number of muscle fibers. This new staging system is intended to be a useful preoperative tool to assess the palatopharyngeal muscle (PPM or posterior pillar). The only way to perform it is during its inspection through the oropharynx during office exams. However, when the tonsils are big, meaning over grade 3, it is impossible to preoperatively assess the PPM, making it best to evaluate it during surgery, especially when repositioning of the muscle is considered.

In recent years, the surgical success rate for treating OSA has increased. This success is related to a better understanding of OSA and the new operative techniques developed specifically and adapted to every patient. All modern pharyngoplasties are concerned with PPM management, which highlights the importance of a better pre-surgical evaluation. Thus, using the PASS may help surgeons determine the best way to treat each OSA patient, highlighting the significance of this study as a way of correlating OSA with PPM anatomy. Further studies are needed to demonstrate the accuracy of the correlation between the PASS and the intrasurgical findings.

5. Conclusions

The PASS has a suitable internal design for evaluating the position of the PPM during oropharyngeal examinations. It is an easy scale to learn and put into practice. The evaluation of the oropharyngeal examination videos here shows that the PASS has excellent interrelationships and moderate internal concordance. This scale may be useful as a common language among sleep surgeons for the management of patients with snoring and/or OSA. The use of the PASS to describe the anatomic structures of the lateral pharyngeal wall may be helpful for selecting the appropriate examination and surgical techniques and should be considered for patients with OSA undergoing surgery because of lateral pharyngeal collapse.

Supplementary Materials: The following supporting information can be downloaded at: https://www.mdpi.com/article/10.3390/life13030709/s1, Videos S1–S5: Five videos of the 23 patients included in the study, explaining the use of the PASS in these patients, and Video S6: One video showing palatopharyngeal muscle dissection during live surgery.

Author Contributions: Original idea, R.L.; design of the analysis, data analysis, and report writing, M.M.M.-B. and M.P.C.-S.; supervision, M.M., G.P., C.O.-R. and R.L.; contributions to the graphics, photographs, and manuscript review, G.B., M.M.M.-B., A.M., N.P.-M., M.M. and G.P. All authors have read and agreed to the published version of the manuscript.

Funding: This research received no external funding.

Institutional Review Board Statement: Institutional review board approval was obtained from Hospital Universitario de Fuenlabrada, Madrid, Spain (IRB): 22/116. Date of approval: 20 December 2022.

Informed Consent Statement: Informed consent was obtained from all subjects in the study.

Data Availability Statement: Not applicable.

Acknowledgments: We thank the entire team, including the nurses that form part of the otorhinolaryngology service in Ronquido Monterrey, and Clínica Hospital Constitución, ISSSTE Monterrey in Mexico, as well as the patients who visited the clinic.

Conflicts of Interest: The authors declare that there is no conflict of interest.

References

1. Benjafield, A.V.; Eastwood, P.R. Sleep apnoea: A literature-based analysis. *Lancet Respir. Med.* **2020**, *7*, 687–698. [CrossRef]
2. Olszewska, E.; Woodson, B.T. Palatal anatomy for sleep apnea surgery. *Laryngoscope Investig. Otolaryngol.* **2019**, *10*, 181–187. [CrossRef]
3. Friedman, M.; Ibrahim, H.; Bass, L. Clinical staging for sleep-disordered breathing. *Otolaryngol. Head Neck Surg.* **2002**, *127*, 13–21. [CrossRef] [PubMed]
4. Friedman, M.; Salapatas, A.M. Updated Friedman staging system for obstructive sleep apnea. *Adv. Otorhinolaryngol.* **2017**, *80*, 41–48.
5. De Vito, A.; Woodson, B.T. OSA upper airways surgery: A targeted approach. *Medicina* **2021**, *57*, 690. [CrossRef] [PubMed]

6. Cammaroto, G.; Bianchi, G. Sleep medicine in otolaryngology units: An international survey. *Sleep Breath.* **2021**, *25*, 2141–2152. [CrossRef] [PubMed]

7. Casale, M.; Moffa, A.; Giorgi, L.; Sabatino, L.; Pierri, M.; Lugo, R.; Baptista, P.; Rinaldi, V. No-cutting remodelling intra-pharyngeal surgery can avoid CPAP in selected OSA patients: Myth or reality? *Eur. Arch. Otorhinolaryngol.* **2022**, *279*, 5039–5045. [CrossRef]

8. Lugo-Saldaña, R.; Saldívar-Ponce, K. *Manual de Procedimientos Quirúrgicos en Ronquido y Apnea Obstructiva del Sueño*; Editorial Amolca S.A.S: Medellín, Colombia, 2022.

9. Okuda, S.; Abe, S.; Kim, H.J.; Agematsu, H.; Mitarashi, S.; Tamatsu, Y.; Ide, Y. Morphologic characteristics of palatopharyngeal muscle. *Dysphagia* **2008**, *23*, 258–266. [CrossRef] [PubMed]

10. Sumida, K.; Ando, Y.; Seki, S.; Yamashita, K.; Fujimura, A.; Baba, O.; Kitamura, S. Anatomical status of the human palatopharyngeal sphincter and its functional implications. *Surg. Radiol. Anat.* **2017**, *39*, 1191–1201. [CrossRef]

11. Lugo-Saldaña, R.; Saldívar-Ponce, K.; González-Sáez, I. Selecting different approaches for palate and pharynx surgery: Palatopharyngeal Arch Staging System. *Int. J. Head Neck Surg.* **2019**, *10*, 92–97. [CrossRef]

12. Cahali, M.B. Lateral pharyngoplasty: A new treatment for OSAHS. *Laryngoscope* **2003**, *113*, 1961–1968. [CrossRef] [PubMed]

13. Pang, K.P.; Woodson, B.T. Expansion sphincter pharyngoplasty: A new technique for the treatment of obstructive sleep apnea. *Otolaryngol. Head Neck Surg.* **2007**, *137*, 110–114. [CrossRef] [PubMed]

14. Korhan, I.; Gode, S.; Midilli, R. The influence of the lateral pharyngeal wall anatomy on snoring and sleep apnea. *J. Park Med. Assoc.* **2015**, *65*, 125–130.

15. Alcaraz, M.; Bosco, G.; Pérez-Martín, N.; Morato, M.; Navarro, A.; Plaza, G. Advanced palate surgery: What works? *Curr. Otorhinolaryngol. Rep.* **2019**, *9*, 271–284. [CrossRef]

16. Friedman, M.; Soans, R.; Gurpinar, B. Interexaminer agreement of Friedman tongue positions for staging of obstructive sleep apnea/hypopnea syndrome. *Otolaryngol. Head Neck Surg.* **2008**, *139*, 372–377. [CrossRef] [PubMed]

17. Yu, J.L.; Rosen, I. Utility of the modified Mallampati grade and Friedman tongue position in the assessment of obstructive sleep apnea. *J. Clin. Sleep Med.* **2020**, *16*, 303–308. [CrossRef]

18. Sundman, J.; Fehrm, J.; Friberg, D. Low inter-examiner agreement of the Friedman staging system indicating limited value in patient selection. *Eur. Arch. Otorhinolaryngol.* **2018**, *275*, 1541–1545. [CrossRef]

19. Sundman, J.; Bring, J.; Friberg, D. Poor interexaminer agreement on Friedman tongue position. *Acta Otolaryngol.* **2017**, *137*, 554–556. [CrossRef]

20. Friedman, M.; Hamilton, C.; Samuelson, C.G.; Lundgren, M.E.; Pott, T. Diagnostic value of the Friedman tongue position and Mallampati classification for obstructive sleep apnea: A meta-analysis. *Otolaryngol. Head Neck Surg.* **2013**, *148*, 540–547. [CrossRef]

21. Ng, S.K.; Lee, D.L.; Li, A.M. Reproducibility of clinical grading of tonsillar size. *Arch. Otolaryngol. Head Neck Surg.* **2010**, *136*, 159–162. [CrossRef]

22. Brodsky, L. Modern assessment of tonsils and adenoids. *Pediatr. Clin. N. Am.* **1989**, *36*, 1551–1569. [CrossRef]

23. Mediano, O.; Mangado, N.G.; Montserrat, J.M.; Alonso-Álvarez, M.L.; Almendros, I.; Alonso-Fernández, A.; Barbé, F.; Borsini, E.; Caballero-Eraso, C.; Cano-Pumarega, I.; et al. International consensus document on obstructive sleep apnea. *Arch. Bronconeumol.* **2022**, *58*, 52–68. [CrossRef] [PubMed]

24. Lu, X.; Zhang, J.; Xiao, S. Correlation between Brodsky Tonsil Scale and tonsil volume in adult patients. *Biomed. Res. Int.* **2018**, *2018*, 6434872. [CrossRef] [PubMed]

25. Friedman, M.; Tanyeri, H.; La Rosa, M. Clinical predictors of obstructive sleep apnea. *Laryngoscope* **1999**, *109*, 1901–1907. [CrossRef] [PubMed]

26. Casale, M.; Moffa, A.; Pignataro, L.; Rinaldi, V. Palatopharyngeus muscle in pharyngoplasty surgery for OSAS: Cut or not to cut? *Eur. Arch. Otorhinolaryngol.* **2021**, *278*, 2657–2658. [CrossRef]

Article

Effects of Rapid Maxillary Expander and Delaire Mask Treatment on Airway Sagittal Dimensions in Pediatric Patients Affected by Class III Malocclusion and Obstructive Sleep Apnea Syndrome

Sara Caruso [1], Emanuela Lisciotto [1,*], Silvia Caruso [1], Alessandra Marino [2], Fabiana Fiasca [1], Marco Buttarazzi [3], David Sarzi Amadè [2], Melania Evangelisti [2], Antonella Mattei [1] and Roberto Gatto [1]

[1] Department MeSVA, University of L'Aquila, 67100 L'Aquila, Italy
[2] Department of Neurosciences, Mental Health and Sensory Organs, Suicide Prevention Center Sant'Andrea Hospital, Sapienza University of Rome, 00185 Roma, Italy
[3] Independent Researchers, 50134 Firenze, Italy
* Correspondence: emanuela.lisciotto@gmail.com

Abstract: Obstructive sleep apnea syndrome (OSAS) is a sleep-related breathing disorder that is very common in pediatric patients. In the literature, there are very few studies concerning the association between OSAS and class III malocclusion in children. The use of a rapid maxillary expander (RME) in association with the Delaire mask is a common treatment protocol for class III malocclusion. The aim of this work was to evaluate the cephalometric variations of upper airway dimensions and OSA-related clinical conditions after orthodontic treatment with an RME and the Delaire mask, as recorded in pediatric patients with a class III malocclusion who were affected by OSAS. In this preliminary study, 14 pediatric patients with mixed dentition, aged between 6 and 10 years, were selected. All patients were treated with an RME and the Delaire mask. Pre- and post-treatment cephalometric radiographs were traced, analyzed, and compared. The results demonstrated a significant increase in the upper airway linear measurements and the nasopharyngeal and oropharyngeal dimensions ($p \leq 0.05$). This increase creates an improvement in airway patency and in OSAS-related clinical conditions. The use of the RME in association with the Delaire mask can be effective in the treatment of pediatric patients with a class III malocclusion who are affected by OSAS.

Keywords: obstructive sleep apnea syndrome; class III malocclusion; Delaire mask; rapid maxillary expansion

1. Introduction

Sleep-related breathing disorders in pediatric patients are caused by partial or complete obstruction of the upper airways, which can cause episodes of hypopnea or apnea with subsequent repercussions on pulmonary ventilation, oxygenation, and sleeping quality [1]. Obstructive sleep apnea syndrome (OSAS) is characterized by day and night symptoms and signs such as snoring, sleep apnea, drowsiness, decreased concentration and memory ability, enuresis, nocturia, increased blood pressure, heart rate changes, night sweats, irritability, dry mouth, and growth retardation. OSAS-related clinical conditions can also accentuate comorbidities in frail pediatric patients [2]. Numerous studies in the scientific literature show that the most frequent cause of OSAS in childhood is represented by adenotonsillar hypertrophy [3] but other risk factors have also been highlighted, such as rhinitis, allergies, snoring, obesity, craniofacial anomalies, and neuromuscular diseases [4,5]. The most frequent craniofacial alterations reported in children with OSAS are retrognathia, midface hypoplasia, contraction of the upper jaw, relative macroglossia, and anterior open bite [6]. Among these craniofacial anomalies, it has been highlighted that hypoplasia of the maxilla and/or mandible may promote the development of sleep-related breathing disorders [7,8]. Two recent systematic reviews evidenced the association between maxillomandibular discrepancy and OSA, suggesting that children with OSA have more skeletal Class II characteristics and a dolichofacial mandibular growth direction compared to normal children [2–4].

The clinical diagnosis of OSAS in children is very sensitive but is not specific to the diagnosis of OSAS when compared to nocturnal polysomnography (PSG) [9]. Clinical history and physical examination only serve the purpose of identifying the subjects in whom clinicians will have to carry out diagnostic instrumental tests [10]. PSG is the current gold standard diagnostic test for OSA, and the apnea-hypopnea index (AHI, which records the number of obstructive and mixed apneas and hypopneas per hour of total sleep time) is the PSG parameter most commonly reported to discriminate the presence and severity of OSA [11]. According to the most recent national guidelines, a diagnosis of OSAS is established when the AHI is >1 ev/h [12]. However, PSG, as the current confirmatory procedure, is cost-prohibitive and is not widely available. Using screening questionnaires for OSA is particularly useful when clinicians need to determine the presence of OSA. Employing specific and accredited questionnaires for the clinical diagnosis of this condition can also offer a good reliability index [11].

Orthopedic therapy is effective in restoring the correct relationship between the maxillary bone and the mandibular bone and between the dental arches. The efficacy of treatment using rapid palate expansion is amply demonstrated in the literature [13–15] in children with OSAS and malocclusion. This treatment can operate on the transversal maxillary deficit and extend the base of the nasal cavities, with an increase in nasopharyngeal and oropharyngeal air space and, consequently, can create an improvement in the patency of the upper respiratory tract [16,17]. In patients with class III malocclusion, the overgrowth of the jaw is often accompanied by hypoplasia of the maxillary bone [18]. Despite this connection, in the literature, there is little evidence of the association between class III malocclusion and sleep-related breathing disorders; therefore, there are no validated therapy protocols regarding this clinical condition.

Therapy with an RME and a Delaire mask is one of the most common orthopedic treatment protocols for Class III malocclusions. Among the variety of treatments proposed, we have adopted a therapeutic protocol for Class III malocclusion by maxillary correction. The excellent results reported in experimental works have influenced the clinical adoption of such an approach [19,20].

The use of both orthopedic devices has a synergistic effect in promoting the growth of the maxilla in the sagittal and transverse dimensions, while simultaneously limiting mandibular overgrowth [21,22].

Nevertheless, very few studies have investigated the use of an RME in association with the protraction of the maxilla to improve the nasopharyngeal and oropharyngeal airway dimensions [23].

Based on these findings, the objective of this work was to evaluate the effects of this protocol treatment via cephalometric radiographs analysis, in terms of upper airway patency, and the improvement of OSAS-related clinical conditions.

2. Materials and Methods

2.1. The Sample

For this study, we selected 14 pediatric patients (6 males and 8 females) with mixed dentition and aged between 6 and 10 years. At the time of diagnosis, all patients had a class III malocclusion and clinical conditions that are associated with OSAS. Before treatment, informed consent was required from the parents of each child. The patients were treated at the Pediatric Sleep Center (Sant'Andrea Hospital, Rome, Italy). The inclusion criteria for the cases were: patients with signs, symptoms, and clinical history of OSAS and PSG diagnosis (mild, moderate, or severe OSAS), patients aged between 6–10 years, with mixed dentition and, in addition, semiotics and cephalometric standards of class III malocclusion.

Moreover, patients with erupted central permanent incisors before the start of orthodontic therapy and those patients who had received treatment with a palate expander according to the same method described below were included. Finally, we included patients for whom lateral teleradiographs were available before and after treatment performed using the same technique, and with whom it was possible to evaluate the patency of the airways.

The exclusion criteria applied to those patients previously treated by surgical therapy for sleep-related breathing disorders (adenoidectomy, tonsillectomy, or adenotonsillectomy) and patients who were taking drugs that could alter respiratory function. Obese patients were also excluded because that may influence the onset of malocclusions [24]. Patients with the characteristics of class I and class II malocclusions, as well as patients with craniofacial anomalies, malformations and/or genetic pathologies, and systemic pathologies, which can alter respiratory functions, were also excluded.

A diagnosis of class III malocclusion was made by clinical examination and confirmed by radiographic evaluation. The patients considered in the study group had clinical conditions associated with sleep-related breathing disorders, a medical history of OSAS, or a previous OSAS diagnosis. Data collection regarding OSAS-related clinical conditions was carried out using clinical examination and specific questionnaires [25]. Afterward, the diagnosis of OSAS was established when the AHI index was >1 ev/h, according to AASM guidelines [12]. For each patient, skeletal, dental, and pharyngeal cephalometric variables were measured and analyzed, then considered on the cephalometric tracings at time T0 before treatment and at T1, at the end of treatment. Cephalometric analysis is useful as a screening test to characterize skeletal and soft tissue relationships in children experiencing OSAS [26]. As a prospective analysis with 95% of power and a level of significance of alpha = 0.05, the sample size was calculated to need at least 9 participants.

For the cephalometric evaluation, both angular and linear variables were measured, as reported in Table 1.

Table 1. Cephalometric variables, as measured and analyzed.

Skeletal Variables	
SNA (grades)	Angle identified by the points S, N, and A
SNB (grades)	Angle identified by the points S, N, and B
ANB (grades)	Difference between SNA and SNB angles
SN^GOGN (grades)	Angle identified by the intersection of the S-N and GO-GN planes
FM^ (grades)	Angle identified by the intersection of the Frankfurt and mandibular planes
MM^ (grades)	Angle identified by the intersection of the bispinal and mandibular planes
Dental variables	
OVB (mm)	Vertical distance between the incisal edge of upper and lower incisor
OVJ (mm)	Horizontal distance between the incisal edge of upper and lower incisor
1^MAX (grades)	Angle identified by the intersection of the axis of the upper incisor and the bispinal plane
1s-N-FH (mm)	Distance between the upper incisor and the N-FH line (Frankfurt plane)
1i-N-FH (mm)	Distance between the lower incisor and the N-FH line (Frankfurt plane)
1i^MAND (grades)	Angle identified by the intersection of the axis of the lower incisor and the mandibular plane
1s-A-Po (mm)	Distance between the upper incisor and the A-Po line
1i-A-Po (mm)	Distance between the lower incisor and the A-Po line
1s^1i (grades)	Angle identified by the intersection of the axes of the upper and lower central incisors
Ul-E-line (mm)	Distance between upper lip (Ul point) and Ricketts's aesthetic line (tip of nose–tip of chin)
Ll-E-line (mm)	Distance between upper lip (Ll point) and Ricketts's aesthetic line (tip of nose–tip of chin)

Table 1. *Cont.*

Upper airway space dimensions	
Nasopharynx (mm)	Distance from a point on the posterior contour of the soft palate to the point closest to the posterior wall of the pharynx
Oropharynx (mm)	Distance from the intersection of the posterior edge of the tongue and the lower edge of the mandible to the point closest to the posterior wall of the pharynx
Hypopharynx (mm)	Distance from a point on the posterior base of the tongue to the point closest to the posterior wall of the pharynx
PNS-P (mm)	Distance between the lowest point of the soft palate (P) and the posterior nasal spine (PNS)
MP-H (mm)	Distance between the mandibular plane (MP) and the hyoid bone (H)
PH L (mm)	Distance between the base of the epiglottis (Eb) and the tip of the tongue (Tt)

Linear measurements are expressed in millimeters (mm); angular measurements (^) are expressed in degrees.

2.2. Orthodontic Assessment and Orthopedic Therapy

All patients were treated with an RME. The rapid palatal expansion devices that were used were fixed, with a central expansion screw and cemented bands on the second deciduous molars and with hooked arms extended to the palatal surfaces of the canines and first deciduous molars. All patients were treated with the same expansion protocol. After cementation, the RME was activated. The rapid expansion protocol performed involved two rounds/day for 15 days. After the expansion time, the RME was blocked and kept in the mouth for an average duration of 12 months. After blocking the RME, patients were treated with a Delaire mask. Extraoral protraction with elastics was performed, directly connected to the cemented RME. Patients had to wear the mask all night and for two hours during the day [21]. The mask treatment had an average duration of 6 months, with a two-month follow-up until the end of the treatment. The overall treatment had an average duration of 18 months.

2.3. Statistical Analysis

Descriptive statistics were compiled to characterize the patient population. Continuous variables (age, dental, and skeletal and oropharyngeal variables) were expressed as median and interquartile ranges (IQRs) since the data were not normally distributed (the Shapiro–Wilk test); categorical variables (gender, clinical conditions associated with respiratory sleep disorders, and the presence of mixed dentition) were expressed as absolute and percentage frequencies. The Wilcoxon signed-rank test was used to compare dental, skeletal, and oropharyngeal variables at T0 and T1. Statistical analyses were performed with STATA/IC 15.1 For all statistical tests, significance was two-sided and was set at $p \leq 0.05$.

3. Results

A total of 14 patients were included in the study with a median age of 8 years old (IQR 7–9). More than half were females ($n = 8/14$, 57%). Before treatment, 64.29% ($n = 9/14$) were affected by apnea, 50% ($n = 7/14$) by rhinitis and allergy, more than half ($n = 8/14$, 57.14%) had adenotonsillar hypertrophy, and only 14.29% ($n = 2/12$) of patients were affected by snoring. No patient was obese. The clinical conditions associated with sleep-related breathing disorders in the study group are summarized in Table 2, showing their frequency in the present sample.

The analysis of the pre-treatment and post-treatment cephalometric variables led to the following results.

At baseline, the median values of the angles 1^MAX, 1^MAND, and 1s^1i were 101° (IQR 95–109) 84.5° (IQR 81–89), and 145° (IQR 139–154), respectively ($p \leq 0.05$). After the orthodontic treatment, the values of the same angles were significantly varied: the first two were increased, 1^MAX 114° (IQR 104–116) and 1^MAND 90° (IQR 87–94), while the third decreased to 1s^1i 131° (IQR123–136). In addition, we found a variation in the sagittal position of the maxilla and mandible in terms of the SNA, SNB, and ANB angles, which varied from 78.5° (IQR 77–81), $p \leq 0.05$), 77° (IQR 76–82), and 2.5° (IQR 1–5) at T0 to 81.5° (IQR 78–87), 77.5° (IQR 75–85), and 3° (IQR 2–4), respectively, at T1. Moreover, before treatment, the median values of 1s-N-FH and 1i-N-FH angles were 41.15 mm (IQR 37.4–45.5) and 39.5 mm (IQR 37–43), respectively; at the end of the treatment, we recorded 44.75 mm (IQR 42–52.5) and 45 mm (IQR 41–49.5), respectively.

A comparison of the dental and skeletal variables analyzed at times T0 and T1 is shown in Table 3.

Table 2. Descriptive characteristics of the sample ($N = 14$).

Gender, n (%)	Male	6 (42.86)
	Female	8 (57.14)
Age, median (IQR)		8 (7–9)
Clinical conditions associated with respiratory sleep disorders, n (%)		
Apnea	No	5 (35.71)
	Yes	9 (64.29)
Rhinitis	No	7 (50.00)
	Yes	7 (50.00)
Roncopathy	No	12 (85.71)
	Yes	2 (14.29)
Andenotonsillar hypertrophy	No	6 (42.86)
	Yes	8 (57.14)
Allergy	No	7 (50.00)
	Yes	7 (50.00)
Mixed dentition, n (%)	No	1 (7.14)
	Yes	13 (92.86)

Variables were expressed as median and interquartile ranges (IQRs).

Table 3. Comparison of the dental and skeletal cephalometric variables at T0 and T1, expressed as median values and IQRs.

	T0 (IQR)	T1 (IQR)	p-Value [#]
SNA (degree)	78.5 (77 81)	81.5 (78 87)	0.041 *
SNB (degree)	77 (76 82)	77.5 (75 83)	0.298
ANB (degree)	2.5 (1 5)	3 (2 4)	0.430
SN^GOGN (degree)	35 (32 38)	34.5 (30 38)	0.173
FM^ (degree)	28 (25 30)	27.5 (24 30)	0.063
MM^ (degree)	27.5 (26.5 32)	29 (27 32)	0.395
1^MAX (degree)	101 (95 109)	114 (104 116)	0.016 *
1s-N-FH (mm)	41.15 (37.4 43.5)	44.75 (42 52.5)	0.014 *
1i-N-FH (mm)	39.5 (37 43)	45 (41 49.5)	0.033 *
1i^MAND (degree)	84.5 (81 89)	90 (87 94)	0.019 *
1s-A-Po (mm)	1.85 (0.2 4.5)	5.5 (2.2 17)	0.028 *
1i-A-Po (mm)	1.9 (1.1 2)	2.15 (2 11)	0.027 *
1s^1i (degree)	145 (139 154)	131 (123 136)	0.020 *
Ls-Le (mm)	2.25 (0 3.9)	1.25 (−2 2)	0.637
Li-Le (mm)	1.5 (0 2.8)	0 (−1.5 1)	0.429
OVB (mm)	1.4 (0 3)	0.85 (0.1 1.5)	0.615
OVJ (mm)	−0.5 (−1.32 3)	2.41 (0.3 3)	0.109

[#] Wilcoxon signed-rank test. * statistical significance, ($p \leq 0.05$). Linear measurements are expressed in millimeters (mm); angular measurements (^) are expressed in degrees.

Several changes in the pharyngeal measurements were also found. At time T0, the median value of the width of the nasopharynx was 7.5 mm (IQR 6–11), while after treatment, it increased to 9.5 mm (IQR 8.4–14, $p \leq 0.05$.) The median value of the oropharyngeal dimension at time T0 was 15.2 mm (IQR 14.3–18), while in T1 it increased to 16.65 mm (IQR 15–19, $p \leq 0.05$.) The hypopharyngeal dimension was also slightly enlarged. The median values relating to the distances PNS-P, MP-H, and PH-L were increased in T1 compared to baseline, with a not statistically significant level ($p > 0.05$). The comparison between the pharyngeal variables measured in T0 and T1 is shown in Table 4. and in Figure 1.

Table 4. Comparison of the oropharyngeal cephalometric variables at T0 and T1, expressed as median values and IQRs.

	T0 (IQR)	T1 (IQR)	*p*-Value [#]
Rinopharynx (mm)	7.5 (6 11)	9.5 (8.4 14)	0.009 *
Oropharynx (mm)	15.2 (14.3 18)	16.65 (15 19)	0.020 *
Hypopharynx (mm)	10.2 (8 14)	13 (10 16)	0.095
PNS-P (mm)	28.5 (25 32)	30 (26 30)	0.875
MP-H (mm)	11.5 (9.35 15)	11 (7.8 19)	0.530
PH-L (mm)	61.2 (54 66)	65 (56.3 70)	0.258

[#] Wilcoxon signed-rank test. * Statistical significance ($p \leq 0.05$). Linear measurements are expressed in millimeters (mm).

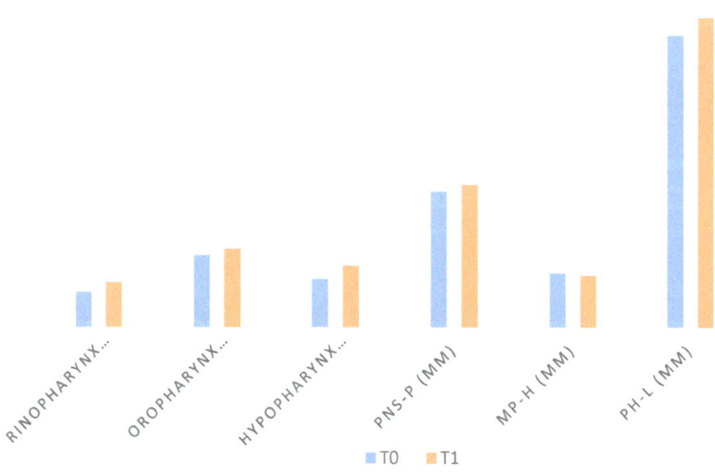

Figure 1. Comparison of the oropharyngeal cephalometric variables at T0 and T1.

4. Discussion

The present investigation analyzed the changes recorded after orthopedic therapy with an RME and a Delaire mask. In the literature, many cephalometric studies have previously evaluated the craniofacial features related to OSAS as predisposing factors in the pathogenesis of upper airway obstruction during sleep [3]. The efficacy of orthopedic treatment with an RME in patients with OSAS was demonstrated in the literature [27–29]. The use of the RME as an orthopedic device is confirmed to be effective in promoting the growth of transverse diameters of the maxillary bone. An increase in the nasopharyngeal and oropharyngeal spaces confirms that treatment with a rapid palatal expander is effective in improving the patency of the upper airways [30–32]. The association between OSAS and class III malocclusion is an infrequent clinical condition. For this reason, there are no validated therapeutic protocols as yet. Orthopedic treatment with an RME and a Delaire mask is widely used in clinical practice to correct the relationship between the maxillary and mandibular bones [20–22]. Among the variety of therapeutic treatments, we chose a protocol for the correction of a Class III malocclusion for the maxilla, thus limiting the approach to Class III cases with maxillary involvement [33,34]. In addition, the positive effects of this treatment on the sagittal pharyngeal dimensions in Class III malocclusion subjects have previously been investigated in the literature [23]. The therapeutic protocol was also chosen in consideration of the fact that mandibular growth has an influence on the size of the upper airways [35–45]. Therefore, it can be hypothesized that jaw growth could also have beneficial effects on the upper airways. Mandibular distraction osteogenesis may also be helpful for treating OSAS in patients with maxillary hypoplasia and severe upper airway obstruction [35].

Based on these findings, the aim of this preliminary study was to confirm the positive effects of an RME and Delaire mask therapy on the upper airway dimensions and evaluate the improvement of OSAS-related clinical conditions. The results obtained from this study showed a significant variation in some of the cephalometric variables analyzed. The increase in SNA angle proves that point A moved anteriorly and suggests a variation in the sagittal position of the maxilla related to the skull base. Bearing in mind that these cephalometric measurements do not usually change in normal conditions, these increments gain importance as they make a contribution to the goals of this therapy; this is one of the attendant effects of the protraction made by the Delaire mask [20,21].

Moreover, no SNB angle decrease was found, as has been described in the study by A.S. Kilinç [23]. Instead, in accordance with M. Rosa's study, no significant changes were found in the sagittal position of the mandible in relation to the skull base (ANB and SNB angles) [21]. This suggests that the correction of class III malocclusion has been determined only BY the forward motion of the maxillary bone and not By below e backward rotation of the jaw.

Regarding the dental parameters, as already described by McNamara and Brudon (1993) [23] and Kim et al. (1999) [35], an increase in the 1^MAX angle was found, which indicates greater uprightness of the upper incisors in relation to the bispinal plane [36–38].

Contrary to what was reported by McNamara, in the current study, there was an increase in the 1^MAND angle and a decrease in the interincisal angle (T0 = 145°, after treatment, T1 = 131°). These findings suggest a correction of the back inclination of the lower incisors. This is also confirmed by changes to the dentition's anterior limit: an increase of 1s-A-Po indicates the greater protrusion of the upper incisor in relation to the basal plane, while an increase of 1i-A-Po indicates greater protrusion of the lower incisor in relation to the basal plane.

Our analysis of airway measurements reported the following results: before treatment, the nasopharyngeal value was 7.5 mm. After the treatment, there was an increase in this value to 9.5 mm. An important result is an increase in oropharyngeal dimension from 15.2 mm, before treatment, to 16.65 mm at the end of the treatment. Therefore, as noted previously by Pavoni et al. [18], significant changes in the parameters related to upper airway patency were observed. A small increase in the median values of the hypopharynx has been noted for the mean values in T0 and T1 of 10.2 mm and 13 mm, respectively, but this occurred with an unequal distribution, and therefore cannot be considered to be statistically significant; also, this value must be considered in relation to the MP-H and PNS-P values [39]. The result was an increase in the median values of the MP-H, PH-L, and PNS-P, which is indicative of a greater distance between the hyoid bone and the mandibular plane and proves an increase in the palate's sagittal dimensions. As already shown in the literature, these changes are closely related to an improvement in OSAS-related clinical conditions [40–42]. In particular, the change in the hyoid bone's position has a significant impact on the shape and position of the tongue, affecting the patency of the airways [18–44]. As reported in other studies, in this study, we found a significant increase in the maxillary forward position. In addition, mandibular forward movement and downward and backward rotation were inhibited; this finding suggests that the most significant changes occurred in the maxillary bone [23]. The use of the Delaire mask for protraction has been proven to be a valid aid in improving nasal breathing [45], which is essential for the correct growth of the palate and the pneumatic development of the maxillary bone and upper airways. These results confirm that a treatment that changes the position of the jaw bones, the tongue, and the soft palate will also have an effect on the oropharyngeal airway dimensions that are closely related to these structures [35,46,47].

Limitations of the Study

In this preliminary study, there is no control group; therefore, it provides no evidence as to the effectiveness of the treatments in varying the cephalometric parameters. Our study is limited to a two-dimensional assessment. No volume ratings were calculated because three-dimensional diagnostic techniques are not a routine form of examination in the pediatric population. The improvement of the clinical conditions in terms of sleep-related breathing disorders occurred preliminarily through a clinical control examination. No values relating to the AHI index at baseline were reported because, in this ongoing study, we aim to perform further evaluations with a polysomnographic follow-up.

5. Conclusions

The current study reports an increase in nasopharyngeal and oropharyngeal spaces. This study's findings confirm that treatment with a rapid maxillary expander, when associated with a Delaire mask, effectively improves the patency of the upper airway's sagittal dimensions. Our preliminary results pave the way for further controlled studies in order to confirm the value of this protocol when used as a treatment in pediatric patients with class III malocclusion who are affected by OSAS.

Author Contributions: Investigation, M.B.; resources, S.C. (Silvia Caruso), A.M. (Alessandra Marino), D.S.A. and M.E.; data curation, A.M. (Antonella Mattei) and F.F.; writing—original draft, E.L.; supervision, S.C. (Sara Caruso) and R.G. All authors have read and agreed to the published version of the manuscript.

Funding: This research received no external funding.

Institutional Review Board Statement: The study was conducted in accordance with the Declaration of Helsinki and was ethically approved by the Institutional Review Board of the University of L'Aquila for studies involving humans (document DR206/2013).

Informed Consent Statement: Informed consent was obtained from both parents of each child involved in the study.

Conflicts of Interest: The authors declare no conflict of interest.

References

1. Katyal, V.; Pamula, Y.; Martin, A.; Daynes, C.; Kennedy, J.; Sampson, W. Craniofacial and upper airway morphology in pediatric sleep-disordered breathing: Systematic review and meta-analysis. *Am. J. Orthod. Dentofac. Orthop.* **2013**, *143*, 20–30. [CrossRef] [PubMed]
2. Pagano, S.; Lombardo, G.; Coniglio, M.; Donnari, S.; Canonico, V.; Antonini, C.; Lomurno, G.; Cianetti, S. Autism spectrum disorder and paediatric dentistry: A narrative overview of intervention strategy and introduction of an innovative technological intervention method. *Eur. J. Paediatr. Dent.* **2022**, *23*, 54–60. [PubMed]
3. Guilleminault, C.; Stoohs, R. Chronic snoring and obstructive sleep apnea syndrome in children. *Lung* **1990**, *168*, 912–919. [CrossRef] [PubMed]
4. Flores-Mir, C.; Korayem, M.; Heo, G.; Witmans, M.; Major, M.; Major, P. Craniofacial morphological characteristics in children with obstructive sleep apnea syndrome. A systematic review and meta-analysis. *J. Am. Dent. Assoc.* **2013**, *144*, 269–277. [CrossRef] [PubMed]

5. Remy, F.; Bonnaure, P.; Moisdon, P.; Burgart, P.; Godio-Raboutet, Y.; Thollon, L.; Guyot, L. Preliminary results on the impact of simultaneous palatal expansion and mandibular advancement on the respiratory status recorded during sleep in OSAS children. *J. Stomatol. Oral Maxillofac. Surg.* **2020**, *892*, 235–240. [CrossRef]
6. Defabianis, P. Impact of nasal airway obstruction on dentofacial development and sleep disturbances in children: Preliminary notes. *J. Clin. Pediatr. Dent.* **2003**, *27*, 95–100. [CrossRef] [PubMed]
7. Ryu, H.-H.; Kim, C.-H.; Cheon, S.-M.; Bae, W.-Y.; Kim, S.-H.; Koo, S.-K.; Ki, M.-S.; Kim, B.-J. The usefulness of cephalometric measurement as a diagnostic tool for obstructive sleep apnea syndrome: A retrospective study. *J. Oral Maxillofac. Surg.* **2015**, *119*, 20–31. [CrossRef]
8. Vale, F.; Albergaria, M.; Carrilho, E.; Francisco, I.; Gumaraes, A.; Caramelo, F.; Malo, L. Efficacy of rapid maxillary expansion in the treatment of obstructive sleep apnea syndrome: A systematic review with meta-analysis. *J. Evid.-Based Dent. Pract.* **2017**, *17*, 159–168. [CrossRef]
9. Caprioglio, A.; Meneghel, M.; Fastuca, R.; Zecca, P.; Nucera, R.; Nosetti, L. Rapid maxillary expansion in growing patients: Correspondence between 3-dimensional airway changes and polysomnography. *Int. J. Pediatr. Otorhinolaryngol.* **2014**, *78*, 23–27. [CrossRef]
10. Goldstein, N.A.; Sculerati, N.; Walsleben, J.A.; Bhatia, N.; Friedman, D.M.; Rapoport, D.M. Clinical diagnosis of pediatric obstructive sleep apnea validated by polysomnography. *Otolaryngol. Head Neck Surg.* **1994**, *111*, 611–617. [CrossRef]
11. Villa, M.P.; Brunetti, L.; Bruni, O.; Cirignotta, F.; Cozza, P.; Donzelli, G.; Ferini-Strambi, L.; Levrini, L.; Mondini, S.; Nespoli, L.; et al. Linee Guida per la diagnosi della Sindrome delle Apnee ostruttive nel sonno in età pediatrica. *Minerva Pediatr.* **2004**, *56*, 239–253. [PubMed]
12. American Academy of Pediatrics. Clinical practice guideline: Diagnosis and management of childhood obstructive sleep apnea syndrome. *Pediatrics* **2002**, *109*, 704–712. [CrossRef] [PubMed]
13. Machado-Júnior, A.-J.; Zancanella, E.; Crespo, A.-N. Rapid maxillary expansion and obstructive sleep apnea: A review and meta-analysis. *Med. Oral Patol. Oral Cirugía Bucal* **2016**, *21*, 465–469. [CrossRef] [PubMed]
14. Villa, M.P.; Rizzoli, A.; Rabasco, J.; Vitelli, O.; Pietropaoli, N.; Cecili, M.; Marino, A.; Malagola, C. Rapid maxillary expansion outcomes in treatment of obstructive sleep apnea in children. *Sleep Med.* **2015**, *16*, 709–716. [CrossRef]
15. Lione, R.; Brunelli, V.; Franchi, L.; Pavoni, C.; Quiroga Souki, B.; Cozza, P. Mandibular response after rapid maxillary expansion in class II growing patients: A pilot randomized controlled trial. *Prog. Orthod.* **2017**, *18*, 36. [CrossRef] [PubMed]
16. Denolf, P.L.; Vanderveken, O.M.; Marklund, M.E.; Braem, M.J. The status of cephalometry in the prediction of non-CPAP treatment outcome in obstructive sleep apnea patients. *Sleep Med. Rev.* **2016**, *27*, 56–73. [CrossRef] [PubMed]
17. Villa, M.P.; Malagola, C.; Pagani, J.; Montesano, M.; Rizzoli, A.; Guilleminault, C.; Ronchetti, R. Rapid maxillary expansion in children with obstructive sleep apnea syndrome: 12-month follow-up. *Sleep Med.* **2007**, *8*, 128–134. [CrossRef]
18. Pavoni, C.; Cretella Lombardo, E.; Lione, R.; Bollero, P.; Ottaviani, F.; Cozza, P. Orthopaedic treatment effects of functional therapy on the sagittal pharyngeal dimensions in subjects with sleep-disordered breathing and Class II malocclusion. *Acta Otorhinolaryngol. Ital.* **2017**, *37*, 479–485. [CrossRef]
19. Keim, R.G.; Gottlieb, E.L.; Nelson, A.H.; Vogels, D.S., 3rd. 2008 JCO study of orthodontic diagnosis and treatment procedures, part 1: Results and trends. *J. Clin. Orthod.* **2008**, *42*, 625–640.
20. McNamara, J.A., Jr.; Brudon, W.L. *Orthodontics and Dentofacial Orthopedics*; Needham Press: Ann Arbor, MI, USA, 2001; pp. 375–385.
21. Rosa, M.; Frumusa, E.; Lucchi, P.; Caprioglio, A. Early Treatment of Class III Malocclusion by RME and Face-Mask Therapy with Deciduous Dentition Anchorage. *Eur. J. Clin. Orthod.* **2013**, *1*, 21–31.
22. Baccetti, T.; Franchi, L.; McNamara, J.A., Jr. Treatment and posttreatment craniofacial changes after rapid maxillary expansion and facemask therapy. *Am. J. Orthod. Dentofac. Orthop.* **2000**, *118*, 404–413. [CrossRef]
23. Kilinç, A.S.; Arslan, S.G.; Kama, J.D.; Ozer, T.; Dari, O. Effects on the sagittal pharyngeal dimensions of protraction and rapid palatal expansion in Class III malocclusion subjects. *Eur. J. Orthod.* **2008**, *30*, 61–66. [CrossRef] [PubMed]
24. Giuca, M.R.; Pasini, M.; Caruso, S.; Tecco, S.; Necozione, S.; Gatto, R. Index of orthodontic treatment need in obese adolescents. *Int. J. Dent.* **2015**, *2015*, 876931. [CrossRef]
25. Brouillette, R.; Hanson, D.; David, R.; Klemka, L.; Szatkowski, A.; Fernbach, S.; Hunt, C. A diagnostic approach to suspected obstructive sleep apnea inchildren. *J. Pediatr.* **1984**, *105*, 10–14. [CrossRef]
26. Battagel, J.M.; Johal, A.; Kotecha, B. A Cephalometric comparison of subjects with snoring and obstructive sleep apnoea. *Eur. J. Orthod.* **2000**, *22*, 353–365. [CrossRef]
27. Marino, A.; Ranieri, R.; Chiarotti, F.; Villa, M.P.; Malagola, C. Rapid maxillary expansion in children with Obstructive Sleep Apnoea Syndrome (OSA). *Eur. J. Paediatr. Dent.* **2012**, *13*, 57–63.
28. Bucci, R.; Montanaro, D.; Rongo, R.; Valletta, R.; Michelotti, A.; D'Antò, V. Effects of maxillary expansion on the upper airways: Evidence from systematic reviews and meta-analyses. *J. Oral Rehabil.* **2019**, *46*, 377–387. [CrossRef] [PubMed]
29. Nota, A.; Caruso, S.; Caruso, S.; Sciarra, F.M.; Marino, A.; Daher, S.; Pittari, L.; Gatto, R.; Tecco, S. Rapid Maxillary Expansion in Pediatric Patients with Sleep-Disordered Breathing: Cephalometric Variations in Upper Airway's Dimension. *Appl. Sci.* **2022**, *12*, 2469. [CrossRef]
30. Haas, A.J. Palatal expansion: Just the beginning of dentofacial orthopedics. *Am. J. Orthod.* **1970**, *57*, 219–255. [CrossRef]

31. Bell, R.A. A review of maxillary expansion in relation to rate of expansion and patient's age. *Am. J. Orthod.* **1982**, *81*, 32–37. [CrossRef]
32. Marino, A.; Malagnino, I.; Ranieri, R.; Villa, M.P.; Malagola, C. Craniofacial morphology in preschool children with obstructive sleep apnoea syndrome. *Eur. J. Paediatr. Dent.* **2009**, *10*, 181–184.
33. Campbell, P.M. The dilemma of Class III treatment: Early or late. *Angle Orthod.* **1983**, *53*, 175–191.
34. Wisth, P.J.; Tritrapunt, A.; Rygh, P.; Boe, O.E.; Nordeval, K. The effect of maxillary protraction on front occlusion and facial morphology. *Acta Odontol. Scand* **1987**, *45*, 227–237. [CrossRef]
35. Kim, J.H.; Viana, M.A.; Graber, T.M.; Omerza, F.F.; BeGole, E.A. The effectiveness of protraction face mask therapy: A meta-analysis. *Am. J. Orthod. Dentofac. Orthop.* **1999**, *115*, 675–685. [CrossRef] [PubMed]
36. Niu, X.; Di Carlo, G.; Cornelis, M.A.; Cattaneo, P.M. Three-dimensional analyses of short- and long-term effects of rapid maxillary expansion on nasal cavity and upper airway: A systematic review and meta-analysis. *Orthod. Craniofac. Res.* **2020**, *23*, 250–276. [CrossRef]
37. Baldini, A.; Nota, A.; Santariello, C.; Caruso, S.; Assi, V.; Ballanti, F.; Gatto, R.; Cozza, P. Sagittal dentoskeletal modifications associated with different activation protocols of rapid maxillary expansion. *Eur. J. Paediatr. Dent.* **2018**, *19*, 151–155.
38. D'Amario, M.; Piccioni, C.; Di Carlo, S.; De Angelis, F.; Caruso, S.; Capogreco, M. Effect of Airborne Particle Abrasion on Microtensile Bond Strength of Total-Etch Adhesives to Human Dentin. *BioMed Res. Int.* **2017**, *2017*, 2432536. [CrossRef] [PubMed]
39. Hoekema, A.; Hovinga, B.; Stegenga, B.; De Bont, L.G. Craniofacial morphology and obstructive sleep apnoea: A cephalometric analysis. *J. Oral Rehabil.* **2003**, *30*, 690–696. [CrossRef]
40. Maltais, F.; Carrier, G.; Cormier, Y.; Sériès, F. Cephalometric measurements in snorers, non-snorers, and patients with sleep apnoea. *Thorax* **1991**, *46*, 419–423. [CrossRef] [PubMed]
41. Mayer, P.; Pépin, J.L.; Bettega, G.; Veale, D.; Ferretti, G.; Deschaux, C.; Lévy, P. Relationship between body mass index, age and upper airway measurements in snorers and sleep apnoea patients. *Eur. Respir J.* **1996**, *9*, 1801–1809. [CrossRef]
42. Ferrazzano, G.F.; Orlando, S.; Cantile, T.; Sangianantoni, G.; Alcidi, B.; Coda, M.; Caruso, S.; Ingenito, A. An experimental in vivo procedure for the standardised assessment of sealants retention over time. *Eur. J. Paediatr. Dent.* **2016**, *17*, 176–180. [PubMed]
43. Turley, P.K. Managing the developing Class III malocclusion with palatal expansion and facemask therapy. *Am. J. Orthod. Dentofac. Orthop.* **2002**, *122*, 349–352. [CrossRef] [PubMed]
44. Yucel, A.; Unlu, M.; Haktanir, A.; Acar, M.; Fidan, F. Evaluation of the upper airway cross-sectional area changes in different degrees of severity of obstructive sleep apnea syndrome: Cephalometric and dynamic CT study. *AJNR Am. J. Neuroradiol.* **2005**, *26*, 2624–2629. [PubMed]
45. Da Silva Filho, O.G.; Magro, A.C.; Capelozza Filho, L. Early treatment of the Class III malocclusion with rapid maxillary expansion and maxillary protraction. *Am. J. Orthod. Dentofac. Orthop.* **1998**, *113*, 196–203. [CrossRef]
46. Sayınsu, K.; Isik, F.; Arun, T. Sagittal airway dimensions following maxillary protraction: A pilot study. *Eur. J. Orthod.* **2006**, *28*, 184–189. [CrossRef] [PubMed]
47. Takada, K.; Petdachai, S.; Sakuda, M. Changes in dentofacial morphology in skeletal Class III children treated by a modified protraction headgear and a chin cup: A longitudinal cephalometric appraisal. *Eur. J. Orthod.* **1993**, *15*, 211–221. [CrossRef]

Article

Association between Patient- and Partner-Reported Sleepiness Using the Epworth Sleepiness Scale in Patients with Obstructive Sleep Apnoea

Konstantinos Chaidas [1],* , **Kallirroi Lamprou [2], John R. Stradling [2] and Annabel H. Nickol [2]**

[1] Ear, Nose, and Throat Department, Oxford University Hospitals NHS Foundation Trust, Oxford OX39DU, UK
[2] Oxford Centre for Respiratory Medicine, Oxford University Hospitals NHS Foundation Trust, Oxford OX37LE, UK
* Correspondence: konchaidas@gmail.com

Abstract: Excessive daytime sleepiness in obstructive sleep apnoea (OSA) is often measured differently by patients and their partners. This study investigated the association between patient- and partner-completed Epworth Sleepiness Scale (ESS) scores and a potential correlation with OSA severity. One hundred two participants, 51 patients and 51 partners, completed the ESS before and three months after initiating CPAP treatment. There was no significant difference when comparing patients' and partners' ESS scores at baseline (10.75 ± 5.29 vs. 11.47 ± 4.96, respectively) and at follow-up (6.04 ± 4.49 vs. 6.41 ± 4.60, respectively). There was a strong correlation between patients' and partners' ESS scores on both (baseline and follow-up) assessments ($p < 0.001$). There was significant improvement in patients' and partners' ESS scores after CPAP therapy ($p < 0.001$). There was no significant difference in patients' or partners' ESS scores between patients with mild, moderate or severe OSA. There was no significant correlation between oxygen desaturation index (ODI) and ESS score reported either by patient or by partner. In conclusion, our study revealed a strong correlation between patient- and partner-reported ESS scores. However, neither patient- nor partner-completed ESS scores were associated with OSA severity.

Keywords: sleepiness; excessive somnolence; Epworth Sleepiness Scale; patient reported Epworth score; partner reported Epworth score; obstructive sleep apnea; continuous positive airway pressure; oxygen desaturation index; bed partner

Citation: Chaidas, K.; Lamprou, K.; Stradling, J.R.; Nickol, A.H. Association between Patient- and Partner-Reported Sleepiness Using the Epworth Sleepiness Scale in Patients with Obstructive Sleep Apnoea. *Life* **2022**, *12*, 1523. https://doi.org/10.3390/life12101523

Academic Editor: Fabrizio Montecucco

Received: 4 September 2022
Accepted: 28 September 2022
Published: 29 September 2022

1. Introduction

Obstructive sleep apnoea (OSA) is an increasingly prevalent sleep disorder affecting 3–7% of men and 2–5% of women [1]. It is characterized by recurrent episodes of partial or complete upper airway collapse during sleep leading to intermittent airflow limitation, sleep fragmentation, arterial oxygen desaturations, and poor sleep quality. The sleep fragmentation and hypoxemia have a variety of medical and functional consequences including cardiovascular and cerebrovascular disease, glucose intolerance, reduced quality of life and daytime sleepiness [2,3]. Excessive daytime sleepiness is considered as a main symptom in OSA and is experienced by most patients [4], but many of them may deny or minimize its degree.

The multiple sleep latency test and the maintenance of wakefulness test are both objective tools to quantify sleepiness, but are time-consuming, laborious and expensive [5]. In order to measure daytime somnolence in common clinical practice, several questionnaires have been developed. One of the simplest and frequently used is the Epworth Sleepiness Scale (ESS) which was first introduced in 1991 and is a tool used to measure the general level of sleepiness in patients with OSA and other sleep disorders in everyday situations [6].

Although ESS has been widely used in both clinical and research settings, there is still controversy in the literature regarding its value as a screening tool for patients with suspected OSA [7–9]. It has been observed that some patients lack insight into the degree of their sleepiness resulting in underestimation of its level. For that reason, many studies attempted to determine if the use of partner-reported ESS score is superior compared to patient- reported ESS score in the evaluation of patients with suspected OSA reporting controversial outcomes [10–12]. Therefore, there is still much confusion regarding the utility of partner-completed ESS in identifying OSA and predicting its severity.

The aim of our study was to assess the association between patient- and partner-completed ESS scores and to investigate a potential correlation between ESS and OSA severity as well as the impact of continuous positive airway pressure (CPAP) therapy on ESS score.

2. Materials and Methods

2.1. Study Protocol

Patients aged over 18 years old accompanied by their partners who were seen at the Oxford Adult Sleep and Ventilation Service Clinic between February 2019 and February 2020, were offered to participate in this prospective study. The study protocol was approved by the Institutional Review Board of Oxford University Hospitals NHS Foundation Trust (Datix ID: 6120).

Participants were excluded from the study if they had previous treatment with CPAP for OSA or prior diagnosis or suspicion of other sleep disorder, if they were unable to complete the ESS or if they had no suitable partner in clinic with them. As "partner" was considered a boyfriend/girlfriend, a spouse or a close relative sharing the same house. Patients without evidence of OSA on the baseline sleep study were also excluded.

After obtaining verbal consent, patients and their partners completed the ESS independently during the initial visit at the clinic. The ESS is a self-administered questionnaire with 8 questions (Figure 1) [6]. Respondents were asked to rate, on a 4-point scale (0–3), their usual chance of dozing off or falling asleep while engaged in eight different activities. The total ESS score (the sum of 8 item scores) can range from 0 to 24. It has been suggested that a cutoff total score of 10 or higher indicates the presence of hypersomnolence.

Use the following scale to choose the most appropriate number for each situation:
0 = would never doze,
1 = slight chance of dozing,
2 = moderate chance of dozing,
3 = high chance of dozing.
It is important that you answer each question as best you can.

Situation: Chance of Dozing (0–3)
1. Sitting and reading ____
2. Watching TV ____
3. Sitting, inactive in a public place (e.g., a theatre or a meeting) __________
4. As a passenger in a car for an hour without a break _________________
5. Lying down to rest in the afternoon when circumstances permit _______
6. Sitting and talking to someone ____________________________________
7. Sitting quietly after lunch without alcohol __________________________
8. In a car, while stopped for a few minutes in the traffic _______________

Figure 1. Epworth Sleepiness Scale.

The following parameters were also obtained from all patients during their first visit in clinic: age, gender, body mass index (BMI), symptoms correlated to OSA, co-morbidities and current medical treatment, smoking history, alcohol consumption, Mallampati grade and neck circumference. After the baseline screening, patients were scheduled to undergo a diagnostic sleep study with a portable machine. The severity of OSA was evaluated based on the oxygen desaturation index (ODI) and the patients were divided into three groups: mild ($5 \leq$ ODI < 15 episodes/h), moderate ($15 \leq$ ODI < 30 episodes/h), and severe (ODI $\geq$ 30 episodes/h) OSA group. CPAP treatment was offered to all patients that had a diagnosis of OSA. Three months after initiating CPAP therapy both patient and partner were again asked to complete the ESS questionnaire independently.

2.2. Statistical Analysis

Data are presented as mean $\pm$ standard deviation, as median (minimum–maximum) or as percentage. The data were tested for normality by using the Kolmogorov–Smirnov test. Changes in outcome variables before and after CPAP therapy were compared by Wilcoxon signed-rank test. Kendall rank correlation test was used to assess the relationship between patient- and partner-reported sleepiness, as well as their association with ODI. Kruskal–Wallis test was used for the comparison between mild, moderate, and severe OSA groups and ESS results. The Bland and Altman method was used to plot the difference between partners' and patients' ESS scores and measure agreement between them. To investigate the presence of a potential trend, linear regression analysis was performed. p values < 0.05 were considered statistically significant. All data were statistically analysed using SPSS software for Windows version 17.0 (SPSS Inc., Chicago, IL, USA).

3. Results

A total of 102 individuals (51 patients and 51 partners) participated in the study and completed the ESS before and three months after initiating CPAP treatment. Table 1 shows baseline characteristics, sleep study results and CPAP usage information for all patients. ESS results obtained from all participants and comparisons between patients' and partners' total and each item's scores are presented in Table 2.

Table 1. Patients' baseline characteristics, sleep study results and CPAP usage information (n = 51).

Variables	Results
Male sex	37 (72.5)
Age, years	60.21 $\pm$ 12.67
Smoking	
Non-smokers	40 (78.4)
Ex-smokers	4 (7.8)
Smokers	7 (13.7)
Comorbidities	
Hypertension	26 (51.0)
Depression	8 (15.7)
Gastroesophageal reflux	7 (13.7)
Diabetes	5 (9.8)
Asthma	4 (7.8)
COPD	3 (5.9)
None	14 (27.5)
BMI, kg/m^2	35.07 $\pm$ 6.23
Neck circumference, cm	42.91 $\pm$ 3.64
Mallampati scale	
Grade 1	11 (21.6)
Grade 2	28 (54.9)

Table 1. *Cont.*

Variables	Results
Grade 3	10 (19.6)
Grade 4	2 (3.9)
ODI, episodes/hour	35.91 ± 23.37
CPAP use, hours/day	4.89 ± 2.46

Values are given as mean ± standard deviation or as number (%). CPAP: continuous positive airway pressure, COPD: Chronic Obstructive Pulmonary Disease, BMI: body mass index, ODI: oxygen desaturation index.

Table 2. Total and each question's ESS scores prior to (baseline) and three months after (follow-up) initiating CPAP therapy.

ESS Score	Patient-Reported		Partner-Reported		ESS Difference (Partner's Minus Patient's Score)	p Value
Baseline	Mean	Median	Mean	Median	Mean	
Total	10.75 ± 5.29	11 (1–19)	11.47 ± 4.96	12 (1–19)	0.72 ± 3.341	0.157
Q1	1.84 ± 1.065	2 (0–3)	1.92 ± 1.036	2 (0–3)	0.08 ± 0.796	0.537
Q2	2.02 ± 0.927	2 (0–3)	2.10 ± 0.922	2 (0–3)	0.08 ± 0.796	0.512
Q3	0.88 ± 0.791	1 (0–3)	0.78 ± 0.808	1 (0–3)	−0.10 ± 0.755	0.354
Q4	1.49 ± 1.102	2 (0–3)	1.71 ± 1.119	2 (0–3)	0.22 ± 1.026	0.131
Q5	2.24 ± 1.012	3 (0–3)	2.35 ± 1.016	3 (0–3)	0.11 ± 1.013	0.335
Q6	0.43 ± 0.608	0 (0–2)	0.43 ± 0.700	0 (0–2)	0.00 ± 0.721	0.980
Q7	1.61 ± 1.133	2 (0–3)	1.94 ± 1.047	2 (0–3)	0.33 ± 0.931	0.013 *
Q8	0.25 ± 0.483	0 (0–2)	0.25 ± 0.523	0 (0–2)	0.00 ± 0.529	1.000
Follow-up						
Total	6.04 ± 4.49	5 (0–19)	6.41 ± 4.60	6 (0–19)	0.37 ± 2.537	0.128
Q1	0.88 ± 0.816	1 (0–3)	1.02 ± 0.905	1 (0–3)	0.14 ± 0.633	0.127
Q2	1.27 ± 0.827	1 (0–3)	1.25 ± 0.891	1 (0–3)	−0.02 ± 0.648	0.827
Q3	0.53 ± 0.758	0 (0–3)	0.49 ± 0.674	0 (0–2)	−0.04 ± 0.631	0.655
Q4	0.65 ± 0.868	0 (0–3)	0.84 ± 0.967	1 (0–3)	0.19 ± 0.917	0.108
Q5	1.37 ± 1.076	1 (0–3)	1.39 ± 1.115	1 (0–3)	0.02 ± 0.707	0.856
Q6	0.24 ± 0.513	0 (0–2)	0.29 ± 0.502	0 (0–2)	0.05 ± 0.465	0.366
Q7	0.94 ± 0.988	1 (0–3)	0.96 ± 0.894	1 (0–3)	0.02 ± 0.735	0.833
Q8	0.16 ± 0.367	0 (0–1)	0.14 ± 0.401	0 (0–2)	−0.02 ± 0.424	0.739

Values are given as mean ± standard deviation or as median (minimum–maximum). ESS: Epworth sleepiness scale, CPAP: continuous positive airway pressure, Q: question. *: $p < 0.05$.

Thirty-seven male and 14 female patients participated in the study. There were no gender-related differences between patient- and partner-completed ESS scores ($p > 0.05$).

3.1. Baseline ESS

A comparison between patient- and partner-reported total ESS score revealed no significant difference (10.75 ± 5.29 vs. 11.47 ± 4.96, respectively, $p = 0.157$). A comparison of the score for each ESS question separately, revealed no significant difference ($p > 0.05$) except for question 7 ($p = 0.013$). Kendall rank correlation revealed significant correlation between patient- and partner-completed ESS baseline scores ($p < 0.001$, Tb = 0.601).

The Bland–Altman plot (Figure 2) shows the individual differences between the two ESS measurements for each patient against the mean ESS score. It did not demonstrate statistically significant differences between partners' and patients' ESS scores ($p = 0.467$, $R^2 = 0.011$).

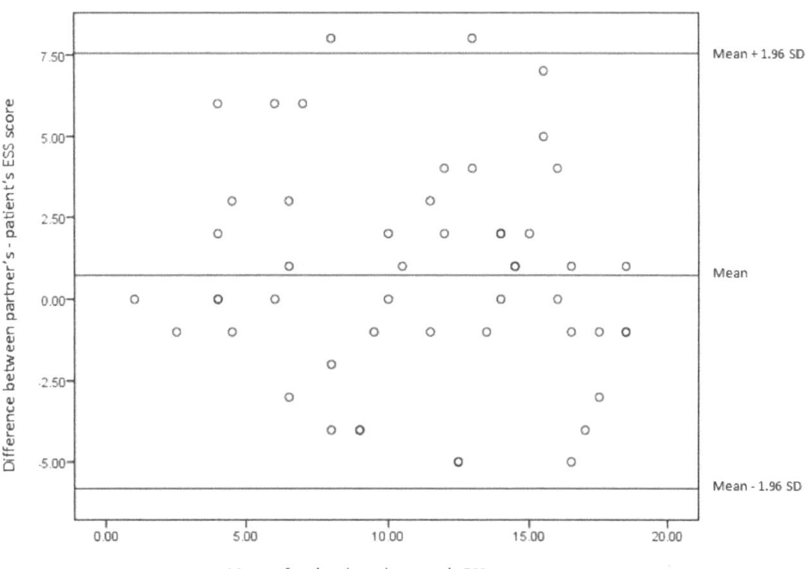

Figure 2. The ESS for paired comparisons is displayed using a Bland–Altman plot. *Y*-axis shows the difference in ESS score (partner's score minus patient's score). *X*-axis shows the mean ESS score (mean score of patient- and partner-reported ESS score). Each dot refers to one patient. The midline indicates the mean difference between the two measurements (partners' minus patients' ESS score). The upper and lower lines indicate the 95% confidence limits for these measures (upper line: mean difference plus $1.96 \times$ SD, lower line: mean difference minus $1.96 \times$ SD, SD: standard deviation).

3.2. Follow-Up ESS

A comparison between patients' and partners' total ESS scores revealed no significant difference (6.04 ± 4.49 vs. 6.41 ± 4.60, respectively, $p = 0.128$). An additional comparison of the score for each ESS question separately revealed no significant difference ($p > 0.05$) for all the questions. Kendall rank correlation test revealed significant correlation between patient- and partner-reported ESS follow-up scores ($p < 0.001$, Tb = 0.651).

3.3. Patients' ESS before and after Treatment

There was statistically significant improvement (reduction) in patient- reported ESS score post CPAP therapy (10.75 ± 5.29 vs. 6.04 ± 4.49, $p < 0.001$). There was a significant reduction in the score for each ESS question ($p < 0.05$) except for question 8 ($p = 0.132$).

3.4. Partners' ESS before and after Treatment

There was statistically significant improvement in partners' ESS score after CPAP treatment (11.47 ± 4.96 vs. 6.41 ± 4.60, $p < 0.001$). There was a significant reduction in the score for each ESS question ($p < 0.05$) except for questions 6 ($p = 0.134$) and 8 ($p = 0.084$).

3.5. Baseline ESS and OSA Severity

Patients were divided into three groups based on OSA severity as determined by ODI (mild, moderate or severe OSA) and Table 3 shows patients' and partners' ESS scores in these groups. There was no significant difference between the groups when comparing the ESS score reported either by patients ($p = 0.534$) or by partners ($p = 0.858$). Likewise, a comparison of patients' ESS score vs. partners' ESS score within the same OSA group revealed no significant difference as shown in Table 3.

Table 3. Baseline (patient- and partner-reported) ESS score in OSA severity groups.

OSA Severity	Number of Patients	Patient-Reported ESS Score		Partner-Reported ESS Score		*p* Value
		Mean	Median	Mean	Median	
Mild	6 (11.8)	12.67 ± 7.50	15.5 (1–19)	12.50 ± 4.68	15 (6–16)	0.786
Moderate	19 (37.3)	10.68 ± 4.84	11 (3–19)	11.53 ± 4.41	11 (5–18)	0.273
Severe	26 (51.0)	10.35 ± 5.18	10.5 (1–19)	11.19 ± 5.53	12 (1–19)	0.253

Values are given as number (%), as mean ± standard deviation or as median (minimum–maximum). ESS: Epworth sleepiness scale, OSA: obstructive sleep apnoea.

Kendall rank correlation test showed no significant correlation between ODI and baseline ESS score reported either by patient ($p = 0.993$, Tb = 0.001) or by partner ($p = 0.794$, Tb = −0.026).

4. Discussion

Our study investigated the association between patient- and partner-reported ESS score demonstrating that a strong correlation exists. Patients and their partners agree in the perception of patient's sleepiness suggesting that the value of ESS in assessing patients for suspected OSA is similar regardless the person completing the questionnaire (patient or partner). On the other hand, it seems that neither patient- nor partner-reported ESS scores are associated with OSA severity and, thus, OSA severity cannot be predicted by ESS alone.

The ESS as a subjective tool to quantify somnolence may be inaccurate in certain situations, especially if the patient is not fully aware of the problem or the partner gives an inaccurately positive estimate of patient's sleepiness. Kumru et al. [10] found a discrepancy between patient's and partner's perception of patient's sleepiness with patients rating their sleepiness lower than their partners. Another study reported that ESS scores of bed partners were higher than those of patients in 67% of the cases suggesting that either the patients tend to underestimate the degree of their sleepiness or their partners overestimate it [12].

In contrast, we demonstrated that partner-completed ESS scores were similar to ESS scores reported by patients. This was also evident by using the Bland–Altman method showing no significant bias between patients' and partners' ESS outcomes. An additional comparison of each ESS item scores revealed no significant difference between patient and partner except for question 7 ("sitting quietly after lunch without alcohol"). Interestingly, this is also the question with the greatest disagreement between patient and partner in the study by Kumru et al. [10]. Although the combination of patient- and partner-completed ESS may help the clinician and increase the accuracy of the results, according to our study findings, the addition of ESS score obtained by the partner does not add much value in the screening process for OSA.

Previous studies have investigated the utility of patient- and partner-completed ESS in predicting OSA severity with contradictory results. Several reports suggest that patients' ESS score does not correlate with OSA severity [8,11,13–16], whereas other studies show that an association is present [12,17–21]. A few studies have also indicated an association between partners' ESS score and OSA severity [12,17,22,23].

A study by Bhat et al. [11] revealed a correlation between partner-completed ESS score and apnoea-hypopnoea index (AHI), but not with ODI or other related parameters. Moreover, the authors found that neither patient- nor partner-completed ESS score alone can predict the severity of OSA. Another study reported that both patients' and partners' ESS scores were independent of AHI levels and, thus, ESS was found to be a poor predictor of AHI [22]. Similarly, Barry et al. [24] demonstrated that the ESS does not correlate with the AHI. In contrast, Walter et al. [12] documented a significant correlation between ESS as estimated by either the patient or the partner and OSA severity. Nevertheless, the authors were unable to identify a cut-off ESS score that correlated strongly with the presence of severe OSA and suggested that their findings may be partially due to the fact that ESS is a subjective questionnaire which can be influenced by a variety of factors such as the quality and duration of sleep the previous night and personality traits.

In light of these contradictory findings, our findings show no correlation between patient- as well as partner-completed ESS and ODI. Therefore, neither patients' nor partners' ESS score can predict OSA severity.

Excessive daytime sleepiness is a common symptom in OSA. Several studies have demonstrated that daytime somnolence levels, both subjective and objective, are not reflective of the severity or even of the presence of OSA [7–9,19,25–27]. This may explain why ESS score may be misleading when used to evaluate patients for OSA. Several factors other than the severity of OSA as demonstrated by sleep study parameters seem to have a significant impact on the degree of daytime sleepiness.

CPAP has been established as the most effective treatment of OSA and has been shown to be associated with a significant reduction in daytime sleepiness in these patients, as well [28,29]. However, on some occasions, sleepiness may persist despite CPAP therapy. In a previous study, 40% of patients with moderate-severe OSA reported sleepiness after three months of treatment with CPAP [30]. Bonsignore et al. showed a prevalence of persistent daytime sleepiness in CPAP-treated patients around 40% in the first three months and 10–20% after that [31]. The same study showed that patients with daily sleepiness at follow-up were younger and more obese, had slightly more severe OSA and were sleepier at baseline. On the other hand, Patel et al. [32] revealed that CPAP therapy significantly reduced the ESS score by a mean of 2.9 points as compared with placebo. On similar lines, another study showed that patients with moderate and severe OSA had a significant improvement in ESS score after one and three months of CPAP therapy [33].

Our results support that patient-completed ESS score, regardless OSA severity, not only significantly improves after 3 months of CPAP therapy, but also normalizes. There was statistically significant reduction in the score for each ESS question except for question 8, which could be explained by the fact that this item is considered as a low soporific situation and, thus, the patients had a low baseline score.

To our knowledge, no other studies have looked at partner-completed ESS after CPAP treatment to assess post therapy patients' daytime sleepiness. We found that there is a statistically significant improvement in partners' ESS score after 3 months of CPAP treatment with a significant reduction in the score for each ESS item except for questions 6 and 8, which are low soporific items. We also found that there was no discrepancy between patient- and partner-completed ESS scores after CPAP treatment.

This study has certain limitations. First, due to the number of participants, conclusions should be made with caution. The patients did not undergo full polysomnography for the diagnosis of OSA. Instead, all patients had a type-3 home sleep study using two respiratory variables, one cardiac variable, one arterial oxygen saturation. Therefore, there were no available data regarding sleep stages, sleep efficiency, number of arousals during sleep, total sleep time, and the presence or not of periodic limb movement (PLM) during sleep. Finally, we used ODI instead of the AHI to make the diagnosis of OSA and classify its severity. However, although AHI is widely used, ODI is as valuable as AHI in diagnosing and grading OSA [34].

5. Conclusions

In contrast with previous studies showing a discrepancy between patients' and partners' ESS scores, our study revealed an association between patient- and partner-reported sleepiness as measured by ESS. In addition, we found that there was no association between ESS score as estimated by either patient or partner and ODI, and hence, OSA severity cannot be predicted by using ESS. Considering the contradictory data in the literature and the growing recognition of daytime sleepiness and its impact on patients' quality of life, more studies are required to investigate the utility of patient- and partner-completed ESS. The ultimate aim should be the development of an accessible, convenient and accurate measure of sleepiness, especially in patients with OSA.

Author Contributions: Conceptualization, K.L. and A.H.N.; methodology, K.L., A.H.N., K.C. and J.R.S.; formal analysis, K.C.; investigation, K.L.; resources, K.L., J.R.S. and A.H.N.; data curation, K.C. and K.L.; writing—original draft preparation, K.C. and K.L; writing—review and editing, K.C., J.R.S. and A.H.N.; visualization, K.C; supervision, J.R.S. and A.H.N.; project administration, K.L., A.H.N. and K.C. All authors have read and agreed to the published version of the manuscript.

Funding: This research received no external funding.

Institutional Review Board Statement: The study was conducted in accordance with the Declaration of Helsinki and approved by the Institutional Review Board of Oxford University Hospitals NHS Foundation Trust (Datix ID: 6120).

Informed Consent Statement: Informed consent was obtained from all subjects involved in the study.

Data Availability Statement: The data presented in this study are available on request from the corresponding author. The data are not publicly available due to privacy related restrictions.

Conflicts of Interest: The authors declare no conflict of interest.

References

1. Punjabi, N.M. The epidemiology of adult obstructive sleep apnea. *Proc. Am. Thorac. Soc.* **2008**, *5*, 136–143. [CrossRef] [PubMed]
2. Moyer, C.A.; Sonnad, S.S.; Garetz, S.L.; Helman, J.I.; Chervin, R.D. Quality of life in obstructive sleep apnea: A systematic review of the literature. *Sleep Med.* **2001**, *2*, 477–491. [CrossRef]
3. Somers, V.K.; White, D.P.; Amin, R.; Abraham, W.T.; Costa, F.; Culebras, A.; Daniels, S.; Floras, J.S.; Hunt, C.E.; Olson, L.J.; et al. Sleep apnea and cardiovascular disease: An American Heart Association/American College of Cardiology Foundation Scientific Statement from the American Heart Association Council for High Blood Pressure Research Professional Education Committee, Council on Clinical Cardiology, Stroke Council, and Council on Cardiovascular Nursing. *J. Am. Coll. Cardiol.* **2008**, *52*, 686–717. [CrossRef] [PubMed]
4. Seneviratne, U.; Puvanendran, K. Excessive daytime sleepiness in obstructive sleep apnea: Prevalence, severity, and predictors. *Sleep Med.* **2004**, *5*, 339–343. [CrossRef]
5. Murray, B.J. Subjective and Objective Assessment of Hypersomnolence. *Sleep Med. Clin.* **2020**, *15*, 167–176. [CrossRef]
6. Johns, M.W. A new method for measuring daytime sleepiness: The Epworth sleepiness scale. *Sleep* **1991**, *14*, 540–545. [CrossRef]
7. Rosenthal, L.D.; Dolan, D.C. The Epworth sleepiness scale in the identification of obstructive sleep apnea. *J. Nerv. Ment. Dis.* **2008**, *196*, 429–431. [CrossRef]
8. Sil, A.; Barr, G. Assessment of predictive ability of Epworth Sleepiness Scale in screening of patients with sleep apnoea. *J. Laryngol. Otol.* **2012**, *126*, 372–379. [CrossRef]
9. Smith, S.S.; Oei, T.P.; Douglas, J.A.; Brown, I.; Jongesen, G.; Andrews, J. Confirmatory factor analysis of the Epworth Sleepiness Scale (ESS) in patients with obstructive sleep apnoea. *Sleep Med.* **2008**, *9*, 739–744. [CrossRef]
10. Kumru, H.; Santamaria, J.; Belcher, R. Variability in the Epworth Sleepiness Scale score between the patient and the partner. *Sleep Med.* **2004**, *5*, 369–371. [CrossRef]
11. Bhat, S.; Upadhyay, H.; DeBari, V.A.; Ahmad, M.; Polos, P.G.; Chokrovery, S. The utility of patient-completed and partner-completed Epworth Sleepiness Scale scores in the evaluation of obstructive sleep apnea. *Sleep Breath.* **2016**, *20*, 1347–1354. [CrossRef] [PubMed]
12. Walter, J.T.; Foldvary, N.; Mascha, E.; Dinner, D.; Golish, J. Comparison of Epworth Sleepiness Scale scores by patients with obstructive sleep apnea and their bed partners. *Sleep Med.* **2002**, *3*, 29–32. [CrossRef]
13. Koehler, U.; Buchholz, C.; Cassel, W.; Hildebrandt, O.; Redhardt, F.; Sohrabi, K.; Töpel, J.; Nell, C.; Grimm, W. Daytime sleepiness in patients with obstructive sleep apnea and severe obesity: Prevalence, predictors, and therapy. *Wien. Klin. Wochenschr.* **2014**, *126*, 619–625. [CrossRef] [PubMed]
14. Lee, S.J.; Kang, H.W.; Lee, L.H. The relationship between the Epworth Sleepiness Scale and polysomnographic parameters in obstructive sleep apnea patients. *Eur. Arch. Otorhinolaryngol.* **2012**, *269*, 1143–1147. [CrossRef] [PubMed]
15. Osman, E.Z.; Osborne, J.; Hill, P.D.; Lee, B.W. The Epworth Sleepiness Scale: Can it be used for sleep apnoea screening among snorers? *Clin. Otolaryngol. Allied Sci.* **1999**, *24*, 239–241. [CrossRef]
16. Dixon, J.B.; Dixon, M.E.; Anderson, M.L.; Schachter, L.; O'brien, P.E. Daytime sleepiness in the obese: Not as simple as obstructive sleep apnea. *Obesity* **2007**, *15*, 2504–2511. [CrossRef]
17. Li, Y.; Zhang, J.; Lei, F.; Liu, H.; Li, Z.; Tang, X. Self-evaluated and close relative-evaluated Epworth Sleepiness Scale vs. multiple sleep latency test in patients with obstructive sleep apnea. *J. Clin. Sleep Med.* **2014**, *10*, 171–176. [CrossRef]

18. Bausmer, U.; Gouveris, H.; Selivanova, O.; Goepel, B.; Mann, W. Correlation of the Epworth Sleepiness Scale with respiratory sleep parameters in patients with sleep-related breathing disorders and upper airway pathology. *Eur. Arch. Otorhinolaryngol.* **2010**, *267*, 1645–1648. [CrossRef]

19. Sforza, E.; Pichot, V.; Martin, M.S.; Barthélémy, J.C.; Roche, F. Prevalence and determinants of subjective sleepiness in healthy elderly with unrecognized obstructive sleep apnea. *Sleep Med.* **2015**, *16*, 981–986. [CrossRef]

20. Damiani, M.F.; Quaranta, V.N.; Falcone, V.A.; Gadaleta, F.; Maiellari, M.; Ranieri, T.; Fanfulla, F.; Carratù, P.; Resta, O. The Epworth Sleepiness Scale: Conventional self vs physician administration. *Chest* **2013**, *143*, 1569–1575. [CrossRef]

21. Manni, R.; Politini, L.; Ratti, M.T.; Tartara, A. Sleepiness in obstructive sleep apnea syndrome and simple snoring evaluated by the Epworth Sleepiness Scale. *J. Sleep Res.* **1999**, *8*, 319–320. [CrossRef] [PubMed]

22. Hughes, C. Can the Epworth sleepiness score predict the apnoea-hypopnoea index in obstructive sleep apnoea and hypopnoea syndrome: A comparitive audit of patient and partner scoring. *Eur. Respir. J.* **2011**, *38*, 3245.

23. Kingshott, R.N.; Sime, P.J.; Engleman, H.M.; Douglas, N.J. Self assessment of daytime sleepiness: Patient versus partner. *Thorax* **1995**, *50*, 994–995. [CrossRef] [PubMed]

24. Barry, A.; Pyne, N.; Munns, S.; Nolan, G. Is the Epworth Sleepiness Scale an appropriate screening tool for patients with sleep disordered breathing (SDB): A retrospective clinical audit. *ERJ Open Res.* **2021**, *7*, 55.

25. Kapur, V.K.; Baldwin, C.M.; Resnick, H.E.; Gottlieb, D.J.; Nieto, F.J. Sleepiness in patients with moderate to severe sleep-disordered breathing. *Sleep* **2005**, *28*, 472–477. [CrossRef]

26. de Castro, J.R.; Rosales-Mayor, E. Clinical and polysomnographic differences between OSAH patients with/without excessive daytime sleepiness. *Sleep Breath.* **2013**, *17*, 1079–1086. [CrossRef]

27. Vgontzas, A.N. Excessive daytime sleepiness in sleep apnea: It is not just apnea hypopnea index. *Sleep Med.* **2008**, *9*, 712–714. [CrossRef]

28. Sullivan, C.E.; Issa, F.G.; Berthon-Jones, M.; Eves, L. Reversal of obstructive sleep apnoea by continuous positive airway pressure applied through the nares. *Lancet* **1981**, *1*, 862–865. [CrossRef]

29. Mason, M.; Welsh, E.J.; Smith, I. Drug therapy for obstructive sleep apnoea in adults. *Cochrane Database Syst. Rev.* **2013**, *31*, CD003002. [CrossRef]

30. Antic, N.A.; Catcheside, P.; Buchan, C.; Hensley, M.; Naughton, M.T.; Rowland, S.; Williamson, B.; Windler, S.; McEvoy, R.D. The effect of CPAP in normalizing daytime sleepiness, quality of life, and neurocognitive function in patients with moderate to severe OSA. *Sleep* **2011**, *34*, 111–119. [CrossRef]

31. Bonsignore, M.R.; Pepin, J.L.; Cibella, F.; Barbera, C.D.; Marrone, O.; Verbraecken, J.; Saaresranta, T.; Basoglu, O.K.; Trakada, G.; Bouloukaki, I.; et al. Excessive Daytime Sleepiness in Obstructive Sleep Apnea Patients Treated With Continuous Positive Airway Pressure: Data From the European Sleep Apnea Database. *Front. Neurol.* **2021**, *12*, 690008. [CrossRef] [PubMed]

32. Patel, S.R.; White, D.P.; Malhotra, A.; Stanchina, M.L.; Ayas, N.T. Continuous positive airway pressure therapy for treating sleepiness in a diverse population with obstructive sleep apnea: Results of a meta-analysis. *Arch. Intern. Med.* **2003**, *163*, 565–571. [CrossRef] [PubMed]

33. Battan, G.; Kumar, S.; Panwar, A.; Atam, V.; Kumar, P.; Gangwar, A.; Roy, U. Effect of CPAP Therapy in Improving Daytime Sleepiness in Indian Patients with Moderate and Severe OSA. *J. Clin. Diagn. Res.* **2016**, *10*, OC14–OC16. [CrossRef] [PubMed]

34. Temirbekov, D.; Gunes, S.; Yazici, Z.M.; Sayin, I. The Ignored Parameter in the Diagnosis of Obstructive Sleep Apnea Syndrome: The Oxygen Desaturation Index. *Turk. Arch. Otorhinolaryngol.* **2018**, *56*, 1–6. [CrossRef]

Article

The Automatic Algorithm of the Auto-CPAP Device as a Tool for the Assessment of the Treatment Efficacy of CPAP in Patients with Moderate and Severe Obstructive Sleep Apnea Syndrome

Beata Brajer-Luftmann [1,*,†], Tomasz Trafas [1,†], Marcin Mardas [2], Marta Stelmach-Mardas [3], Halina Batura-Gabryel [1] and Tomasz Piorunek [1]

1   Department of Pulmonology, Allergology and Pulmonary Oncology, Poznan University of Medical Sciences, Szamarzewskiego 84 Street, 60-569 Poznan, Poland
2   Department of Ginecological Oncology, Institute of Oncology, Poznan University of Medical Sciences, Szamarzewskiego 84 Street, 61-569 Poznan, Poland
3   Department of Treatment of Obesity, Metabolic Disorders and Clinical Dietetics, Poznan University of Medical Sciences, Szamarzewskiego 84 Street, 61-569 Poznan, Poland
*   Correspondence: bbrajer@ump.edu.pl; Tel.: +48-61-841-70-61
†   These authors contributed equally to this work.

Abstract: Obstructive sleep apnea syndrome (OSAS) is a common sleep-related breathing disorder where precise treatment assessment is of high importance. We aimed to validate an automatic algorithm of the auto-CPAP device and reveal polygraph usefulness in the OSAS diagnosis and treatment of outpatients. One hundred patients with moderate OSAS, severe OSAS, and excessive daytime sleepiness qualified for CPAP treatment were included. The study was conducted in three stages. The first stage included a minimum 6-hour polysomnographic examination to select moderate and severe OSAS. The second stage involved an auto-CPAP treatment lasting at least 4 h with simultaneous polygraph recording. The third stage was a titration of at least 4 h with auto-CPAP. The Apnea–Hypopnea Index (AHI) and oxygen desaturation index (ODI) were calculated under auto-CPAP treatment, simultaneously using polygraph (stage two), and as a result of treatment with auto-CPAP (stage three). The mean AHI was 40.0 ± 20.9 for OSAS. Auto-CPAP treatment was effective in 97.5%. The mean residual AHI was 8.6 ± 4.8; there was no significant difference between the AHI CPAP, and the AHI polygraph values were assessed with an accuracy of 3.94/h. The sensitivity and specificity of calculated cut point 8.2 event/hour were: 55% and 82%, respectively. The calculated AUC for the AHI CPAP parameter was 0.633. Presented data confirmed that the automatic algorithm of auto-CPAP is a good tool for the assessment of the treatment efficacy of CPAP in patients, i.e., home setting, with a moderate or severe stage OSAS-presented high sleepiness.

Keywords: OSAS; polysomnography; polygraphy; CPAP

Citation: Brajer-Luftmann, B.; Trafas, T.; Mardas, M.; Stelmach-Mardas, M.; Batura-Gabryel, H.; Piorunek, T. The Automatic Algorithm of the Auto-CPAP Device as a Tool for the Assessment of the Treatment Efficacy of CPAP in Patients with Moderate and Severe Obstructive Sleep Apnea Syndrome. *Life* **2022**, *12*, 1357. https://doi.org/10.3390/life12091357

Academic Editor: Konstantinos Chaidas

Received: 31 July 2022
Accepted: 25 August 2022
Published: 31 August 2022

1. Introduction

Obstructive sleep apnea syndrome (OSAS) is a common, but often unrecognized, disorder [1]. It was estimated that nearly one billion adults aged 30–69 years worldwide could have obstructive sleep apnea. The number of people with moderate to severe obstructive sleep apnea, for which treatment is generally recommended, is estimated to be almost 425 million [2]. It is a disorder characterized by the repeated complete (apnea) or partial (hypopnea) narrowing of the upper respiratory tract with the work of the respiratory muscles preserved [3–5]. The consequence of OSAS is the deterioration of blood oxygenation and frequent (usually unconscious) awakenings, leading to defragmentation of sleep and excessive daytime sleepiness, which leads to an increased prevalence of cardiovascular diseases in patients with OSAS [3–8]. In our previous study, we confirmed that lifestyle modification resulting in the reduction of one unit of body mass index (BMI) gives meaningful and positive changes in selected cardio-metabolic risk factors, such as total cholesterol (TC), triglycerides (TG), fasting insulin, and blood pressure (BP), in OSAS patients [9].

However, the gold standard treatment for OSAS is continuous positive airway pressure (CPAP) therapy [10]. Titration with the CPAP device is a method that relies on the continuous generating of overpressure in the upper airways to prevent their collapse [11]. The standard is the selection of therapeutic pressure through a simultaneous polysomnographic recording during which the patient is treated with a CPAP device. The method of selecting pressure depends on the type and frequency of respiratory disorders [12]. CPAP treatment is recommended in severe disease with Apnea–Hypopnea Index (AHI) > thirty, moderate disease with AHI > fifteen and severe daytime sleepiness assessed on Epworth Sleepiness Scale (ESS) $\geq$ eleven points or cardiovascular complications, and mild disease of five $\leq$ AHI $\leq$ fifteen with severe daytime sleepiness [4,13]. According to the American Academy of Sleep Medicine (AASM) guidelines, the attempted therapy should take place in a sleep laboratory [14]. In Poland and also in poor-income countries, the availability of specialized laboratories is still insufficient, both at the stage of OSAS diagnosis and treatment implementation [15]. The situation is made worse by the lack of outpatient procedures. There seems to be a justified need for research aimed at demonstrating the usefulness of outpatient devices in the diagnosis and treatment of OSAS.

The presented work is an attempt to validate the automatic algorithm of the auto-CPAP device to reduce breathing disorders during sleep in patients with OSAS. In this way, the authors tried to check that the automatic algorithm of the auto-CPAP treatment may be enough for OSAS treatment effectiveness assessment in a situation of limited access to polysomnography (PSG), especially in primary care.

2. Materials and Methods

2.1. Study Design

This is a prospective observational study. Patients were recruited between July 2019 and June 2020 at the Department of Pulmonology, Allergology, and Pulmonary Oncology, Poznan University of Medical Sciences (Poland). All included patients were diagnosed and treated in the University Sleep Laboratory. All procedures were conducted to good laboratory and diagnostic practices.

The study protocol was approved by the Bioethical Committee at Poznan University of Medical Science 7 March 2019 (No: 339/19). All enrolled participants provided written informed consent. The study was conducted in accordance with the Helsinki Declaration.

2.2. Study Population

One hundred adult patients (fifty men and fifty women) were included in the study according to the following criteria: age over 18, moderate and severe OSAS diagnosed, and associated excessive daytime sleepiness qualified for treatment with CPAP. The exclusion criteria were as follow: the lack of excessive daytime sleepiness; impaired patency of the upper respiratory tract resulting in reduced effectiveness of positive pressure therapy—indications for ear, nose, and throat (ENT) intervention, taking hypnotic, and/or sedatives.

Excessive daytime sleepiness was assessed with the Epworth Sleepiness Scale (ESS) and defined as achieving $\geq$ 11 points [16]. The diagnosis of OSAS was performed according to AASM recommendations [14,17] using a polysomnography machine (Alice 6, Philips, PA, USA, 2017).

2.3. Auto-CPAP Device Validation

The study was carried out in three independent stages, performed during a stay in the Sleep Laboratory for 2 days.

The first stage included a minimum 6-hour polysomnographic examination (Polysomnograph Alice 6, Murrysville, PA, USA, 2017) with a particular assessment of the following parameters: AHI, oxygen desaturation index (ODI), average saturation, and stages of sleep (NREM-N1, N2, N3, and REM). The analysis was performed with the Sleepware G3 software (Respironics, Murrysville, PA, USA, 2019).

The second stage involved treatment with an auto-CPAP lasting at least 4 h with simultaneous polygraphy (PG) recording. In this stage, the Alice Night One polygraph (Philips, PA, USA, 2017) and auto-CPAP Dreamstation (Philips, PA, USA, 2017) were used with a built-in automatic algorithm to maintain the patency of the upper respiratory tract. The values of AHI, ODI, average saturation, therapeutic pressure representing the 90th percentile of pressures generated during therapy, and average pressure generated with the auto-CPAP device were recorded. To distinguish the parameters used in analyses for this study the AHI indicator from the polygraph's test has been modified into an abbreviation AHI PG. The desaturation index from the PG study was designated ODI PG. The analysis was performed with Sleepware software (Respironics, PA, USA, 2019).

The third stage of the study was a titration of at least 4 h with the auto-CPAP apparatus taking into account the following parameters: AHI residual respiratory disorder index, 90th percentile therapeutic pressure generated over time pressure therapy, and the average pressure generated during treatment with the auto-CPAP device. Each patient was provided with a comfortable and tight mask that allowed free breathing. To distinguish the parameters used in the analyses for this study, the AHI indicator from auto-CPAP titration was modified to the abbreviation AHI CPAP. The Encore Basic (Respironics, PA, USA, 2019) software was provided by the auto-CPAP manufacturer. According to the manufacturer's information, auto-CPAP responds to respiratory events in accordance with the principles of manual titration.

2.4. Statistical Analysis

A population of 105 subjects was required to show differences with type I error stated as alpha 0.05 and with the power of 95%. The data were expressed as mean, median, and standard deviation. The normality of the distribution was checked by the Shapiro–Wilk test. The Mann–Whitney test was used to analyze the differences between the variables from the diagnostic stage and auto-CPAP therapy and to describe the differences before and after medical intervention. Using the Spearman's R correlation the relationship between AHI and ODI indices from the therapeutic process was demonstrated. The validation of the automatic algorithm of the auto-CPAP apparatus was made using the Bland– Altman chart and a mountain plot. A p-value < 0.05 defined statistically significant differences. All calculations were performed with the use of Statistica 10 software (TIBICO Software Inc., Palo Alto, CA, USA, 2017).

3. Results

One hundred five adult patients were recruited for the study. Five subjects did not meet inclusion criteria (indications for (ENT) intervention—three patients, and taking sedatives—two patients)—Figure 1.

The study included 100 (50% men) patients diagnosed with at least moderate OSAS and associated excessive daytime sleepiness (ESS $\geq$ 11 points) who qualified for treatment with CPAP air prosthesis. The basic features of the study population and the clinical characteristics of patients with moderate and severe OSAS are presented in Table 1.

The results of the Mann–Whitney U Test ($p = 0.5126$), with the adopted significance level ($\alpha = 0.05$), indicate statistically significant differences between the distributions of variables from the diagnostic stage (PSG) and after the CPAP treatment process. On the other hand, the results of the Mann–Whitney U Test ($p < 0.0001$) indicate no significant differences between the AHI values derived from CPAP and polygraph (PG). Spearman's rank correlation ($R = 0.7193$, $p < 0.0001$) shows a significant relationship between AHI CPAP and AHI PG (Figure 2).

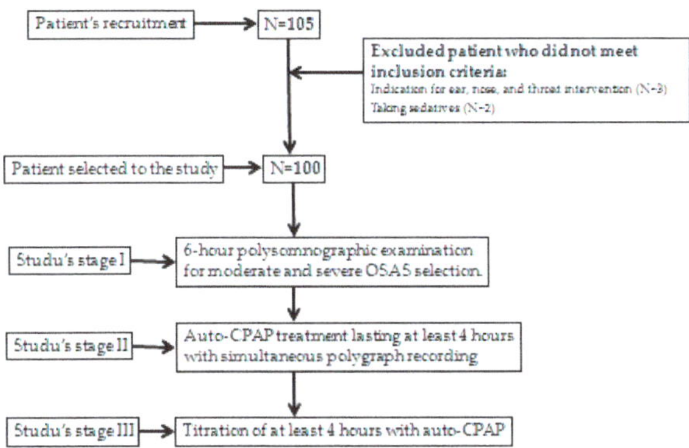

Figure 1. The patient sample selection.

Table 1. CPAP treatment results analyzed using auto-CPAP with polygraph.

Analyzed Variable	Mean	Minimum	Maximum	SD	Lower Quartile	Upper Quartile	Median	*p*-Value
AHI CPAP (1/h)	8.6	1.1	22.3	4.8	4.9	10.9	7.6	<0.0001
AHI PG (1/h)	8.3	1.0	23.0	4.6	4.5	10.8	7.5	<0.0001
ODI PG (1/h)	9.4	2.1	39.3	6.2	5.8	10.7	7.9	<0.0001
Mean saturation PG (%)	92.3	74.0	96.0	3.6	91.0	95.0	3.6	<0.0001
Lowest saturation PG (%)	83.7	58.0	93.0	6.5	81.0	88.5	6.5	<0.0001

Abbreviations: AHI CPAP, Apnea–Hypopnea Index analyzed by auto-CPAP; AHI PG, Apnea–Hypopnea Index scored by polygraph; ODI PG, oxygen desaturation index scored by polygraphy; PG, polygraphy.

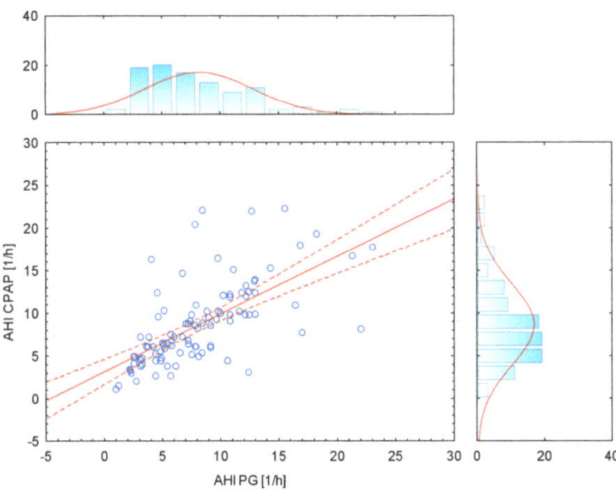

Figure 2. Relationship between AHI CPAP and AHI PG.

In order to validate the operation of the automatic algorithm of the CPAP device, statistical methods were used to compare the AHI residual index values from the titration during CPAP titration and the AHI and ODI values from the manual interpretation of the polygraphy (AHI PG and ODI PG).

A significant correlation was demonstrated between the AHI parameters analyzed by CPAP and scored in polygraphy (Figure 2). However, a significant relationship was demonstrated between the parameters of AHI analyzed by CPAP and the ODI scored in polygraphy (Table 2).

Table 2. The relationship between AHI analyzed by CPAP and the ODI scored in polygraphy.

Analyzed Variable	R Spearman	*p*-Value
AHI CPAP and AHI PG	0.7193	<0.0001
AHI CPAP and ODI PG	0.4435	<0.0001

Abbreviations: AHI CPAP, Apnea–Hypopnea Index analyzed by auto-CPAP; AHI PG, Apnea–Hypopnea Index scored in polygraphy; ODI PG, oxygen desaturation index scored in polygraphy; PG, polygraphy.

Differences between the results of the residual respiratory disorder index obtained from individual patients by means of apparatus titration auto-CPAP and the results from the Alice Night One polygraph were compared with the mean value of the results obtained from both test methods. The solid blue line represents the error systematic and the dashed red lines indicate the 95% compliance limits (Figure 3).

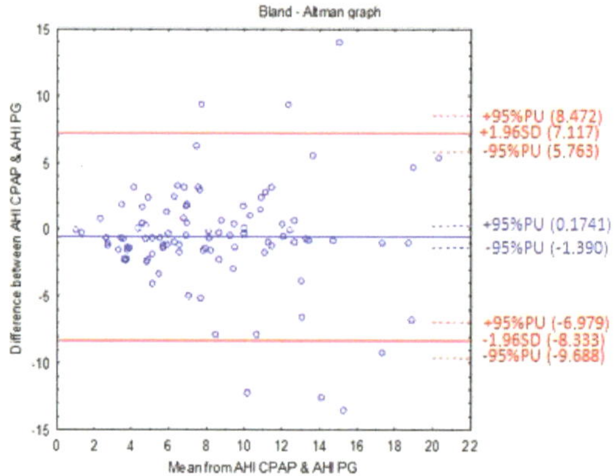

Figure 3. The mean difference between the AHI CPAP and AHI PG. The data are presented using Bland-Altman graph.

The mean difference between the AHI CPAP and AHI PG values was 0.6080 l/h with an accuracy of 3.94 1/h (Table 3).

Table 3. The mean difference between the AHI CPAP and AHI PG.

Analyzed Variable	Mean Difference	SD	Standard Error
Difference between AHI CPAP and AHI PG	0.6080	3.94	0.39

These data are presented using Bland–Altman analysis. Abbreviations: AHI CPAP, Apnea–Hypopnea Index analyzed by auto-CPAP; AHI PG, Apnea–Hypopnea Index scored in polygraphy.

To confirm the identity of both methods, a mountain plot was made, which is a graphical representation of the differences in the values of AHI CPAP residual indices and ODI PG relative to the expected value. A strong positive correlation between the AHI PSG and ODI PSG values (Figure 4) was used for the preparation mountain plot.

By analogy, the value of the AHI PG index has been established as the expected value for the residual respiratory AHI CPAP disturbance indices and ODI PG. The chart is created by determining the percentiles for the ranked ones ascending differences between the AHI CPAP and ODI PG values relative to the value expected AHI PG. A slight shift in the maximum peaks is noticeable as it curves from the zero point, which proves the high compliance with the expected value. The shape of the curves confirms a little discrepancy in the determination of the residual index breathing disorders between both methods.

The determined ROC curves show the sensitivity and specificity of the AHI parameter PSG (Figure 5) and the cut-off point above which the value of the residual index AHI CPAP is most consistent with the real value. The sensitivity and specificity of the calculated cut point 8.2 1/h are 55% and 82%, respectively. The calculated AUC for the AHI CPAP parameter is 0.633.

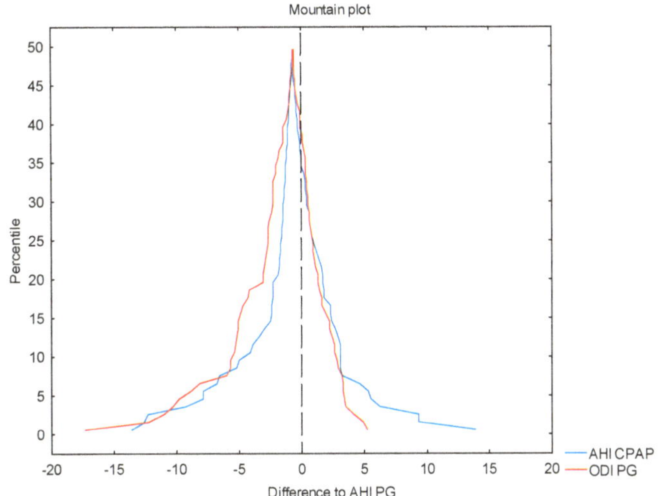

Figure 4. The correlation between the AHI PSG and ODI PSG values (mountain plot).

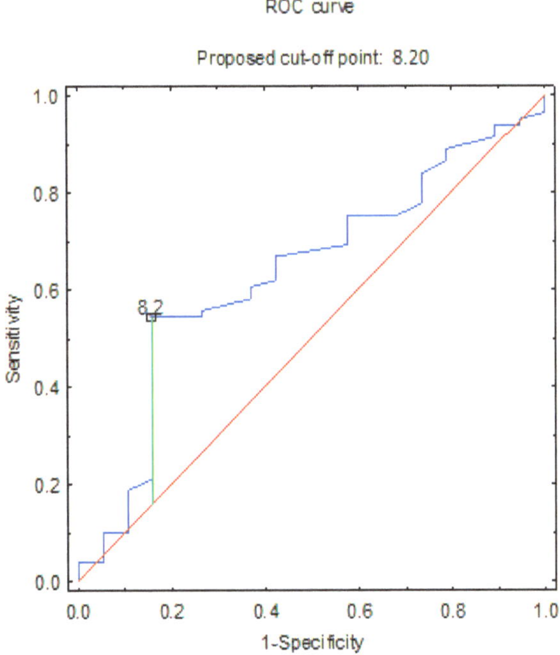

Figure 5. The sensitivity and specificity of the AHI parameter PSG and the cut-off point above which the value of the residual index AHI CPAP is most consistent with the real value using ROC curve.

4. Discussion

The current study is an attempt to validate the automatic algorithm of the auto-CPAP device in the reduction in breathing disorders during OSAS treatment. According to the current AASM guidelines [17], therapy with the CPAP device should be monitored using a polysomnographic device. The insufficient number of polysomnographic laboratories in Poland (currently OSAS diagnostics can be performed in 57 canters), the long waiting time for the examination, and the lack of reimbursed diagnostic procedures performed on an outpatient basis were the reasons for the author to look for a reliable diagnostic and therapeutic method that could improve the existing situation. The possibility of titration with an auto-CPAP device was analyzed without needing to use devices to supervise the operation of an air prosthesis and to evaluate the value of residual AHI. The obtained results indicate a 97.5% effectiveness of AHI reduction as a result of auto-CPAP therapy. The results of the current study suggest that in the case of limited access to PSG, an automatic algorithm of the auto-CPAP seems a sufficient tool to assess the effectiveness of OSAS treatment.

Objective assessment of this issue required a careful selection of patients participating in this project. Corral et al. [18] showed a high agreement of the AHI respiratory disturbance index values obtained from type one and type three devices, according to the AASM guidelines [19]. It has been shown that the sensitivity of the polygraph increases with the value of the AHI respiratory disturbance index [20]. Taking into account the effectiveness of polygraph diagnostics and the poor availability of polysomnography, other authors decided to use a polygraph to evaluate the titration of the auto-CPAP apparatus [21,22].

The current study results revealed a high effectiveness of CPAP treatment using polygraphy and automatic CPAP algorithm. Similar effectiveness of positive pressure therapy was demonstrated by Kotzian et al. [19], pointing to the groups of patients for whom a correction of the therapeutic pressure value is required to enable an effective reduction in AHI. Immediate reduction in daytime sleepiness was confirmed in the majority of patients, distinguishing at the same time a group for which the effect of one night was not sufficient [19]. Similar observations were made by Djonlagic et al. [20]. The authors of this study validated the effectiveness of reducing respiratory disturbances by an automatic algorithm for recognizing respiratory events in which the auto-CPAP device is equipped. Some of the performed statistical analyses showed high compliance of the residual index of respiratory disorders from auto-CPAP and the monitoring device polygraph. Spearman's rank correlation showed a significant relationship between the AHI CPAP and AHI PG values and a clear relationship between AHI CPAP and ODI PG. Gagnadoux et al. [23] confirmed these results. The Passing–Bablok regression analysis carried out by Gagnadoux et al. [23] showed the identity of both research methods. According to the interpretation of the statistical method used, auto-CPAP titration, without the use of surveillance equipment, can be used to determine the residual AHI value, which is a cheaper and simpler method of initial treatment.

Our study showed that titration performed only with the auto-CPAP algorithm is sufficient for the proper course of OSAS therapy in patients with moderate and severe disease, excessive daytime sleepiness, and no indications of ENT treatment. The results of research works published by Corral [18], Botokeky [24], and Nigro [25] confirm these observations. Auto-CPAP titration after the first month of treatment is more effective than manual inpatient titration in patients requiring positive pressure therapy and without serious comorbidities. This results in better adherence to the recommendations for CPAP therapy, a noticeable increase in the number of patients using long-term CPAP, and a reduced percentage of patients discontinuing treatment [20]. Observation of the therapy at home by partners has a positive effect on patient's compliance with treatment with positive blood pressure [22,24]. On the other hand, over the years, psychological measures of behavior change constructs have been increasingly recognized as the most consistent predictors of CPAP adherence and, as a result, the most successful interventions to optimize adherence have been behavioral. Combining theory-based behavioral approaches with telemedicine technology could be the answer to increasing CPAP adherence rates in the real world, although randomized trials are still needed and socioeconomic barriers to telemedicine will need to be addressed to promote health equity [26].

The Bland–Altman difference diagram used in the current study showed a high agreement of the obtained residual AHI values from both methods. The mean difference in the obtained measurements was 0.6080 events per hour. One of the criteria for enrolling patients in the study was the diagnosis of excessive daytime sleepiness. Effective therapy with positive pressure, objectively confirmed in the measurements carried out, also allows the subjective feeling of the effects of treatment, among others in the form of relief from sleepiness. Noseda et al. [27], in his work, showed that in patients with excessive daytime sleepiness, titration with auto-CPAP is as effective as manual titration with polysomnography and CPAP prosthesis, and the obtained therapeutic pressures do not differ significantly. In our study, we confirmed the equivalence of both methods in determining the residual AHI, which was presented graphically by means of a slope chart. The shape of the obtained graph, as well as a slight shift of the peaks in relation to the zero value, indicate a high compatibility of both methods. The presented study indicates no statistically significant differences between the distributions of AHI CPAP and AHI PG. The results of our analysis are confirmed in the previously conducted research by Li et al. [28]. In purpose to show statistical differences between the values of the parameters AHI, ODI, mean saturation, and the lowest saturation, this study was supplemented with a non-parametric analysis of variance test, which showed that the variables from two study stages do not belong to one population. Herkenrath et al. [29] showed similar conclusions, showing statistically significant differences between the medians of the AHI index during diagnosis and treatment with the auto-CPAP device.

5. Limitations

This study has a few limitations. It was conducted only in one center. The study group consisted of patients with moderate to severe OSAS and high sleepiness, creating a relatively small sample size. This is a limitation of the possibility of extrapolating the obtained results to the general population of patients with OSAS. In the patients included in the study, we noticed problems with the proper subjective assessment of sleepiness. Therefore, it was eliminated mainly by collecting an interview from accompanying persons—sleep partners. A certain inconvenience of the study is the failure to obtain data distribution in accordance with the normal distribution, resulting, inter alia, from the non-random selection of patients and the numerical predominance of people with the severe stage of the disease.

6. Conclusions

In conclusion, our study results suggest that the automatic algorithm of the auto-CPAP device is a good tool for the assessment of the treatment efficacy of CPAP in patients with a moderate or severe stage OSAS-presented high sleepiness. This method of assessing the effectiveness of CPAP treatment can be used especially in primary care and in countries where access to PSG is still insufficient.

Author Contributions: Conceptualization, T.T., B.B.-L. and T.P.; methodology, T.T., B.B.-L. and T.P.; software, T.T.; validation, M.M. and M.S.-M.; formal analysis, B.B.-L., M.M. and T.P.; investigation, T.T.; resources, H.B.-G.; data curation, B.B.-L.; writing—original draft preparation, B.B.-L., T.T. and M.S.-M.; writing—review and editing, B.B.-L. and M.S.-M.; visualization, T.T. and M.M.; supervision, T.P.; project administration, M.M.; funding acquisition, H.B.-G. All authors have read and agreed to the published version of the manuscript.

Funding: This research received no external funding.

Institutional Review Board Statement: The study was conducted in accordance with the Declaration of Helsinki and approved 7 March 2019 by the Bioethical Committee at Poznan University of Medical Science (No: 339/19).

Informed Consent Statement: Informed consent was obtained from all subjects involved in the study.

Data Availability Statement: To obtain access to secondary data, please contact corresponding author.

Acknowledgments: The authors thank all the participants for their cooperation.

Conflicts of Interest: The authors declare no conflict of interest.

References

1. Lyons, M.M.; Bhatt, N.Y.; Pack, A.I.; Magalang, U.J. Global burden of sleep-disordered breathing and its implications. *Respirology* **2020**, *25*, 690–702. [CrossRef]
2. Benjafield, A.V.; Ayas, N.T.; Eastwood, P.R.; Heinzer, R.; Ip, M.S.M.; Morrell, M.J.; Nunez, C.M.; Patel, S.R.; Penzel, T.; Pépin, J.L.; et al. Estimation of the global prevalence and burden of obstructive sleep apnoea: A literature-based analysis. *Lancet Respir. Med.* **2019**, *7*, 687–698. [CrossRef]
3. Kushida, C.A.; Littner, M.R.; Morgenthaler, T.; Alessi, C.A.; Bailey, D.; Coleman, J., Jr.; Friedman, L.; Hirshkowitz, M.; Kapen, S.; Kramer, M.; et al. Practice parameters for the indications for polysomnography and related procedures: An update for 2005. *Sleep* **2005**, *28*, 499–521. [CrossRef] [PubMed]
4. Berry, R.B.; Brooks, R.; Gamaldo, C.; Harding, S.M.; Lloyd, R.M.; Quan, S.F.; Troester, M.T.; Vaughn, B.V. AASM Scoring Manual Updates for 2017 (Version 2.4). *J. Clin. Sleep Med.* **2017**, *15*, 665–666. [CrossRef] [PubMed]
5. Epstein, L.J.; Kristo, D.; Strollo, P.J., Jr.; Friedman, N.; Malhotra, A.; Patil, S.P.; Ramar, K.; Rogers, R.; Schwab, R.J.; Weaver, E.M.; et al. Adult Obstructive Sleep Apnea Task Force of the American Academy of Sleep Medicine. Clinical guideline for the evaluation, management and long-term care of obstructive sleep apnea in adults. *J. Clin. Sleep Med.* **2009**, *15*, 263–276.
6. Liu, A.; Cardell, J.; Ariel, D.; Lamendola, C.; Abbasi, F.; Kim, S.H.; Holmes, T.H.; Tomasso, V.; Mojaddidi, H.; Grove, K.; et al. Abnormalities of Lipoprotein Concentrations in Obstructive Sleep Apnea Are Related to Insulin Resistance. *Sleep* **2015**, *38*, 793–799. [CrossRef]
7. Abbasi, A.; Gupta, S.S.; Sabharwal, N.; Meghrajani, V.; Sharma, S.; Kamholz, S.; Kupfer, Y. A comprehensive review of obstructive sleep apnea. *Sleep Sci.* **2021**, *14*, 142–154. [CrossRef]
8. Lal, C.; Weaver, T.E.; Bae, C.J.; Strohl, K.P. Excessive Daytime Sleepiness in Obstructive Sleep Apnea. Mechanisms and Clinical Management. *Ann. Am. Thorac. Soc.* **2021**, *18*, 757–768. [CrossRef]
9. Stelmach-Mardas, M.; Brajer-Luftmann, B.; Kuśnierczak, M.; Batura-Gabryel, H.; Piorunek, T.; Mardas, M. Body Mass Index Reduction and Selected Cardiometabolic Risk Factors in Obstructive Sleep Apnea: Meta-Analysis. *J. Clin. Med.* **2021**, *10*, 1485. [CrossRef]
10. Randerath, W.; Verbraecken, J.; de Raaff, C.A.L.; Hedner, J.; Herkenrath, S.; Hohenhorst, W.; Jakob, T.; Marrone, O.; Marklund, M.; McNicholas, W.T.; et al. European Respiratory Society guideline on non-CPAP therapies for obstructive sleep apnoea. *Eur. Respir. Rev.* **2021**, *30*, 210200. [CrossRef]
11. Boyer, L.; Philippe, C.; Covali-Noroc, A.; Dalloz, M.-A.; Rouvel-Tallec, A.; Maillard, D.; Stoica, M.; d'Ortho, M.P. OSA treatment with CPAP: Randomized crossover study comparing tolerance and efficacy with and without humidification by ThermoSmart. *Clin. Respir. J.* **2019**, *13*, 384–390. [CrossRef] [PubMed]
12. Montserrat, J.M.; Ferrer, M.; Hernandez, L.; Farre, R.; Vilagut, G.; Navajas, D.; Badia, J.R.; Carrasco, E.; De Pablo, J.; Ballester, E.; et al. Effectiveness of CPAP treatment in daytime function in sleep apnea syndrome: A randomized controlled study with an optimized placebo. *Am. J. Respir. Crit. Care Med.* **2001**, *164*, 608–613. [CrossRef] [PubMed]
13. Kikuchi, A.; Sakamoto, K.; Sato, K.; Nakashima, T.; Hashimoto, T.; Hara, H.; Nagura, M. Combination therapy for severe sleep apnea syndrome. *Nihon Jibiinkoka Gakkai Kaiho* **2005**, *108*, 787–793. [CrossRef] [PubMed]
14. Patil, S.P.; Ayappa, I.A.; Caples, S.M.; Kimoff, R.J.; Patel, S.R.; Harrod, C.G. Treatment of Adult Obstructive Sleep Apnea with Positive Airway Pressure: An American Academy of Sleep Medicine Clinical Practice Guideline. *J. Clin. Sleep Med.* **2019**, *15*, 335–343. [CrossRef] [PubMed]
15. Pływaczewski, R. *Częstość i Nasilenie Zaburzeń Oddychania w Czasie snu Wśród Dorosłej Populacji Prawobrzeżnej Warszawy*; Praca habilitacyjna; Instytut Gruźlicy i Chorób Płuc w Warszawie: Warsaw, Poland, 2003.
16. Borsini, E.; Blanco, M.; Schonfeld, S.; Ernst, G.; Salvado, A. Performance of Epworth Sleepiness Scale and tiredness symptom used with simplified diagnostic tests for the identification of sleep apnea. *Sleep Sci.* **2019**, *12*, 287–294. [CrossRef]
17. Wahba, N.; Sayeeduddin, S.; Diaz-Abad, M.; Scharf, S.M. The utility of current criteria for split-night polysomnography for predicting CPAP eligibility. *Sleep Breath.* **2019**, *23*, 729–734. [CrossRef]
18. Corral, J.; Sánchez-Quiroga, M.Á.; Carmona-Bernal, C.; Sánchez-Armengol, Á.; de la Torre, A.S.; Durán-Cantolla, J.; Egea, C.J.; Salord, N.; Monasterio, C.; Terán, J.; et al. Spanish Sleep Network. Conventional Polysomnography Is Not Necessary for the Management of Most Patients with Suspected Obstructive Sleep Apnea. Noninferiority, Randomized Controlled Trial. *Am. J. Respir. Crit. Care. Med.* **2017**, *196*, 1181–1190. [CrossRef]
19. Kotzian, S.T.; Schwarzinger, A.; Haider, S.; Saletu, B.; Spatt, J.; Saletu, M.T. Home polygraphic recording with telemedicine monitoring for diagnosis and treatment of sleep apnoea in stroke (HOPES Study): Study protocol for a single-blind, randomised controlled trial. *BMJ Open* **2018**, *8*, e018847. [CrossRef]
20. Djonlagic, I.; Guo, M.; Matteis, P.; Carusona, A.; Stickgold, R.; Malhotra, A. First night of CPAP: Impact on memory consolidation attention and subjective experience. *Sleep Med.* **2015**, *16*, 697–702. [CrossRef]

21. Cundrle, I.J.; Belehrad, M.; Jelinek, M.; Olson, L.J.; Ludka, O.; Sramek, V. The utility of perioperative polygraphy in the diagnosis of obstructive sleep apnea. *Sleep Med.* **2016**, *25*, 151–155. [CrossRef]

22. Tedeschi, E.; Carratu, P.; Damiani, M.F.; Ventura, V.A.; Drigo, R.; Enzo, E.; Ferraresso, A.; Sasso, G.; Zambotto, F.M.; Resta, O. Home unattended portable monitoring and automatic CPAP titration in patients with high risk for moderate to severe obstructive sleep apnea. *Respir. Care* **2013**, *58*, 1178–1183. [CrossRef] [PubMed]

23. Gagnadoux, F.; Pevernagie, D.; Jennum, P.; Lon, N.; Loiodice, C.; Tamisier, R.; van Mierlo, P.; Trzepizur, W.; Neddermann, M.; Machleit, A.; et al. Validation of the System One RemStar Auto A-Flex for Obstructive Sleep Apnea Treatment and Detection of Residual Apnea-Hypopnea Index: A European Randomized Trial. *J. Clin. Sleep Med.* **2017**, *13*, 283–290. [CrossRef] [PubMed]

24. Botokeky, E.; Freymond, N.; Gormand, F.; Le Cam, P.; Chatte, G.; Kuntz, J.; Liegeon, M.N.; Gaillot-Drevon, M.; Massardier-Pilonchery, A.; Fiquemont, A.; et al. Benefit of continuous positive airway pressure on work quality in patients with severe obstructive sleep apnea. *Sleep Breath.* **2019**, *23*, 753–759. [CrossRef] [PubMed]

25. Nigro, C.A.; Borsini, E.; Dibur, E.; Larrateguy, L.; Cazaux, A.; Elias, C.; de la Vega, M.; Berrozpe, C.; Maggi, S.; Grandval, S.; et al. Indication of CPAP without a sleep study in patients with high pretest probability of obstructive sleep apnea. *Sleep Breath.* **2020**, *24*, 1043–1050. [CrossRef]

26. Bakker, J.; Weaver, T.E.; Parthasarathy, S.; Aloia, M.S. Adherence to CPAP: What Should We Be Aiming For, and How Can We Get There? *Chest* **2019**, *155*, 1272–1287. [CrossRef]

27. Noseda, A.; Andre, S.; Potmans, V.; Kentos, M.; de Maertelaer, V.; Hoffmann, G. CPAP with algorithm-based versus titrated pressure: A randomized study. *Sleep Med.* **2009**, *10*, 988–992. [CrossRef] [PubMed]

28. Li, Q.Y.; Berry, R.B.; Goetting, M.G.; Staley, B.; Soto-Calderon, H.; Tsai, S.C.; Jasko, J.G.; Pack, A.I.; Kuna, S.T. Detection of upper airway status and respiratory events by a current generation positive airway pressure device. *Sleep* **2015**, *38*, 597–605. [CrossRef]

29. Herkenrath, S.D.; Treml, M.; Anduleit, N.; Richter, K.; Pietzke-Calcagnile, A.; Schwaibold, M.; Schäfer, R.; Alshut, R.; Grimm, A.; Hagmeyer, L.; et al. Extended evaluation of the efficacy of a proactive forced oscillation technique-based auto-CPAP algorithm. *Sleep Breath.* **2020**, *24*, 825–833. [CrossRef]

Communication

Improvements in Plasma Tumor Necrosis Factor-Alpha Levels after a Weight-Loss Lifestyle Intervention in Patients with Obstructive Sleep Apnea

Michael Georgoulis [1,*], **Nikos Yiannakouris** [1], **Roxane Tenta** [1], **Ioanna Kechribari** [1], **Kallirroi Lamprou** [2], **Emmanouil Vagiakis** [2] and **Meropi D. Kontogianni** [1]

[1] Department of Nutrition and Dietetics, School of Health Sciences and Education, Harokopio University of Athens, 70 Eleftheriou Venizelou Str., 17676 Athens, Greece

[2] Center of Sleep Disorders, 1st Department of Critical Care, Evangelismos General Hospital, 45-47 Ipsilantou Str., 10676 Athens, Greece

* Correspondence: mihalis.georgoulis@gmail.com; Tel.: +30-210-9549359

Abstract: Obstructive sleep apnea (OSA) and systemic inflammation typically coexist within a vicious cycle. This study aimed at exploring the effectiveness of a weight-loss lifestyle intervention in reducing plasma tumor necrosis factor-alpha (TNF-a), a well-established modulator of systematic inflammation in OSA. Eighty-four adult, overweight patients with a diagnosis of moderate-to-severe OSA were randomized to a standard care (SCG, $n = 42$) or a Mediterranean lifestyle group (MLG, $n = 42$). Both groups were prescribed continuous positive airway pressure (CPAP), while the MLG additionally participated in a 6-month behavioral intervention aiming at healthier weight and lifestyle habits according to the Mediterranean pattern. Plasma TNF-a was measured by an immunoenzymatic method both pre- and post-intervention. Drop-out rates were 33% ($n = 14$) for the SCG and 24% ($n = 10$) for the MLG. Intention-to-treat analysis ($n = 84$) revealed a significant decrease in median TNF-a only in the MLG (from 2.92 to 2.00 pg/mL, $p = 0.001$). Compared to the SCG, the MLG exhibited lower follow-up TNF-a levels (mean difference adjusted for age, sex, baseline TNF-a and CPAP use: -0.97 pg/mL, $p = 0.014$), and further controlling for weight loss did not attenuate this difference ($p = 0.020$). Per protocol analyses ($n = 60$) revealed similar results. In conclusion, a healthy lifestyle intervention can lower plasma TNF-a levels in patients with OSA.

Keywords: sleep-disordered breathing; inflammation; tumor necrosis factor alpha; obstructive sleep apnea; weight loss; dietary intervention; lifestyle intervention; Mediterranean diet

Citation: Georgoulis, M.; Yiannakouris, N.; Tenta, R.; Kechribari, I.; Lamprou, K.; Vagiakis, E.; Kontogianni, M.D. Improvements in Plasma Tumor Necrosis Factor-Alpha Levels after a Weight-Loss Lifestyle Intervention in Patients with Obstructive Sleep Apnea. *Life* **2022**, *12*, 1252. https://doi.org/10.3390/life12081252

Academic Editor: Adrian Neagu

Received: 14 July 2022
Accepted: 13 August 2022
Published: 17 August 2022

Publisher's Note: MDPI stays neutral with regard to jurisdictional claims in published maps and institutional affiliations.

1. Introduction

Obstructive sleep apnea (OSA) and systemic inflammation typically coexist and combinedly contribute to increased cardiovascular risk [1]. Continuous positive airway pressure (CPAP) is currently the first line treatment for OSA, but its effectiveness in improving patients' inflammatory profile remains questionable [1]. Although weight loss and the adoption of a healthy lifestyle may reduce inflammation [2], only a few interventional studies have explored the anti-inflammatory benefits of lifestyle modification in patients with OSA, showing improvements in C-reactive protein (CRP) levels [3,4]. Besides CRP, tumor necrosis factor-alpha (TNF-a) is a well-established modulator of systematic inflammation and has been proposed as a key biomarker for the onset and progression of OSA; chronic intermittent hypoxia, a key characteristic of respiratory pathology in OSA, can induce TNF-a expression and lead to cardiovascular complications, while TNF-a inhibition has been shown to ameliorate OSA progression [5].

In previous reports of the MIMOSA (Mediterranean diet/lifestyle Intervention for the Management of Obstructive Sleep Apnea) randomized controlled clinical trial [6–8], we demonstrated that the combination of a 6-month weight-loss dietary/lifestyle intervention based on the Mediterranean pattern with the prescription of CPAP therapy can lead to greater improvements in OSA patients' polysomnographic indices, symptoms and cardiometabolic manifestations, including reductions in high-sensitivity CRP, compared to CPAP alone; we additionally showed that a Mediterranean lifestyle intervention, a holistic approach for beneficial dietary, physical activity and sleep modification according to the Mediterranean pattern, is superior in improving several sleep-related, cardiometabolic and oxidative stress indices in patient with OSA compared to the Mediterranean diet per se. In the present study we aimed at further exploring the potential benefits of this weight-loss Mediterranean lifestyle intervention in ameliorating systemic inflammation in OSA through improvements in plasma TNF-a levels.

2. Materials and Methods

This manuscript presents secondary analyses of the MIMOSA trial (registered at ClinicalTrials.gov, identifier: NCT02515357) in a subsample with available plasma TNF-a measurements. The study protocol has been previously presented in detail [6–8] and was approved by the Ethics Committee of Harokopio University. The study population consisted of 84 newly-diagnosed, adult, overweight male and female patients with a polysomnography (PSG)-extracted apnea-hypopnea index (AHI) $\geq$ 15 events/h (indicative of moderate-to-severe OSA), who provided a signed written consent and were randomly allocated to a standard care group (SCG, n = 42) or a Mediterranean lifestyle group (MLG, n = 42). Patients of both groups were prescribed CPAP therapy as the standard care for OSA management. Specifically, following the initial diagnostic PSG, participants were subjected to an overnight in-laboratory CPAP titration sleep study and were accordingly prescribed the same auto-CPAP device on the basis of current recommendations for OSA management [9,10]. Patients were asked to obtain the prescribed auto-CPAP device, were given detailed information on CPAP technical standards, cleaning and maintenance and were instructed to use it daily during night sleep for the whole study period.

On top of CPAP, patients in the MLG also participated in a 6-month lifestyle modification program aiming at a healthier body weight, improving dietary habits towards a Mediterranean-style diet (i.e., a dietary pattern characterized by abundance of fruits, vegetables, non-refined grains, legumes, nuts and seeds; olive oil as the principal fat source; moderate consumption of dairy products, white meat and fish/seafood; prudent alcohol intake; limited amounts of red meat and the avoidance of processed foods, such as sweets, sugar-sweetened beverages and fast food) [11], the adoption of a physically active lifestyle ($\geq$150 min/week of any kind of lifestyle physical activity or organized exercise) [12] and optimal night-time sleep duration (7–9 h/day) [13]. The lifestyle intervention was structured in seven, 60-min, group (3–5 patients) counselling sessions led by the research dietitian, performed biweekly for the first two months and then monthly for the next four months of the study. In brief, the first session was devoted to weight loss and emphasis was given to dietary practices that can help reduce energy intake, such as food portion control, correct identification of hunger and satiety and proper meal conditions. Patients were also provided with pedometers and were asked to record their total daily steps, aiming at a gradual increase with the ultimate goal of 10,000 steps/day. In the following six sessions, patients were gradually trained to increase adherence to the principles of the Mediterranean lifestyle. In each session patients were informed about the nutritional value and health effects of specific food groups and were given goals about their optimal consumption according to the Mediterranean diet pyramid [11]. Other healthy dietary/lifestyle practices, such as ensuring a nutritional variety, choosing unprocessed, traditional, local and seasonal Mediterranean foods; implementing healthy cooking techniques; adopting a physically active lifestyle with emphasis on outdoor convivial activities; and sleep hygiene were also addressed. The intervention was based on cognitive-behavioral therapy, with emphasis on goal setting, problem solving, self-monitoring of lifestyle habits (patients were asked to record specific lifestyle parameters in self-monitoring print forms on a daily basis throughout the 6-month intervention, such as the consumption of major food groups, the duration of physical activity and the duration of night-sleep, in order to evaluate adherence to intervention goals and enhance motivation), stimulus control, managing high-risk situations and relapse prevention to facilitate behavioral change [14,15]. At the beginning of each session, the research dietitian weighed the patients, reviewed their self-monitoring forms and patients were asked to report difficulties/barriers in achieving the lifestyle goals set at the previous session. Then, patients were led in a group discussion to create a plan for dealing with the difficulties/barriers described. In the second part of each session, new lifestyle goals were set, the adherence to which was evaluated in the next session.

Participants' anthropometric indices, lifestyle habits and 12-h fasting plasma TNF-a levels were assessed pre- and post-intervention. Body weight (kg) and height (m) were measured following a standardized protocol; the body mass index (BMI) was calculated as weight divided by height squared, and participants were classified as overweight or obese according to international BMI cut-off points. Dietary habits, in terms of habitual food/food group consumption were evaluated through a food frequency questionnaire previously validated in the adult Greek population [16], and adherence to the Mediterranean diet was evaluated through the Mediterranean Diet Score (MedDietScore) [17]. The MedDietScore is an a priori index, taking into account the habitual consumption of nine food groups (i.e., whole grains, potatoes, fruits, vegetables, legumes, full-fat dairy products, fish, poultry, and red meat and products), olive oil and alcohol. Based on the recommendations of the Mediterranean diet, the consumption of each of the 11 components of the index is scored using a scale that ranges from 0 to 5. For foods typical of the Mediterranean diet, i.e., whole grains, potatoes, fruits, vegetables, legumes, fish and olive oil, scoring ranges from 0 to 5 for a very rare to a very frequent consumption, respectively, while the opposite scale (i.e., 0 for a very frequent to 5 for a very rare consumption) is used for foods not typically consumed in the Mediterranean diet, i.e., full-fat dairy products, poultry and red meat. Alcohol consumption was given a score of 0 for no consumption or consumption of >7 standardized servings per day, and scores of 1 to 5 for the consumption of 6–7, 5–6, 4–5, 3–4 and <3 standardized servings per day, respectively (1 standardized serving equals to 12 g of ethanol). The total MedDietScore ranges from 0 to 55, with higher values indicating a greater level of adherence to the Mediterranean diet. The short-form of the International Physical Activity Questionnaire [18] was used to evaluate physical activity habits, and total daily time (min/day) of physical activity was calculated for each participant. Daily duration (h/day) of night sleep and CPAP use were self-reported by participants. TNF-a levels were measured by an immunoenzymatic method (Human TNF-alpha Quantikine ELISA Kit, R&D Systems, Minneapolis, MN, USA); the intra-assay and inter-assay variation coefficients were <5% and <8%, respectively.

The MIMOSA trial was originally powered to detect a significant effect size of the lifestyle intervention on the AHI in a population of 180 patients. For the purpose of the present secondary analysis, post hoc power calculation revealed that the study subsample (n = 84) was sufficient to obtain ≥80% power to detect a difference in follow-up TNF-a levels between the SCG and the MLG, allowing for a type-I error rate of 0.05. The intention-to-treat method [19] was applied in the primary analyses and a secondary per protocol analysis was also performed. Analyses were conducted using the SPSS software version 23 (IBM Corp. 2015, Armonk, NY, USA) and p-values < 0.050 indicated statistically significant results. The Shapiro–Wilk test was used to assess the normality of continuous variables. Differences between groups were tested through the chi-square test for categorical variables or the Student's t-test and the Mann–Whitney U test for normal and skewed continuous variables, respectively. Changes from baseline within each group and differences between groups in plasma TNF-a were tested through the Wilcoxon signed-rank test and the Mann–Whitney U test, respectively. The analysis of covariance was applied to explore mean differences (MD) and 95% confidence intervals (CI) between groups in TNF-a at the 6-month follow-up; age, sex, baseline values of TNF-a, CPAP use (h/day), %Δ weight [((follow-up weight − baseline weight)/baseline weight) × 100] and follow-up AHI levels served as covariates.

3. Results

The trial flow diagram of the MIMOSA study can be found in previously published reports [6–8]. For the needs of the present secondary analyses, 84 newly-diagnosed patients with moderate-to-severe OSA originally randomized in two groups (n = 42 in the SCG and n = 42 in the MLG) and for whom plasma TNF-a measurements were available, consisted the final study population. Of those, 24 patients were lost to follow-up, 14 (33%) in the SCG and 10 (24%) in the MLG. The main reasons for study discontinuation in the MLG were circumstances or events that made participation in the counselling sessions unfeasible, the lack of interest in the intervention, group-session scheduling conflicts or the unjustified complete loss of contact with the research dietitian. Participants in the SCG were considered dropouts if they did not complete the 6-month re-evaluation (e.g., could not be reached to schedule an appointment, refused to participate in the follow-up due to lack of time or interest or did not show up for their scheduled appointment).

The baseline characteristics of the study population are shown in Table 1. Mean age was 47 ± 9 years, males accounted for 81% of the study sample and obesity prevalence was 82%. Participants exhibited a moderate adherence to the Mediterranean diet (mean MedDietScore: 31.9 ± 4.6), low physical activity level [median (1st, 3rd quartile) min/day: 12.9 (4.29, 34.3)] and an inadequate mean daily night-time sleep duration (6.1 ± 1.5 h/day, compared to the recommended 7–9 h/day for adults). The median (1st, 3rd quartile) AHI value was 58.0 (26.0, 89.0) events/h and 74% of participants had an AHI $\geq$ 30 events/h, indicative of severe disease. Although all enrolled patients were prescribed with CPAP, 86% of the SCG and 79% of the MLG acquired the device and started the treatment at baseline (p = 0.393). Among users, mean daily CPAP use was 4.32 ± 2.45 h/day in the SCG and 3.41 ± 2.45 h/day in the MLG (p = 0.128). No significant differences between the SCG and the MLG were observed in baseline sociodemographic, lifestyle and clinical characteristics.

Table 1. Baseline characteristics of the study population.

	Total (n = 84)	SCG (n = 42)	MLG (n = 42)	p [a]
Age, years	46.5 ± 9.4	46.5 ± 9.4	46.5 ± 9.5	0.991
Male sex, n (%)	68 (81)	35 (83)	33 (79)	0.578
BMI, kg/m^2	35.5 ± 5.5	35.6 ± 5.4	35.5 ± 5.7	0.953
Obesity, n (%) [b]	69 (82)	34 (81)	35 (83)	0.776
MedDietScore (0–55) [c]	31.9 ± 4.6	32.1 ± 4.6	31.7 ± 4.7	0.709
Physical activity, min/day	12.9 (4.29, 34.3)	10.7 (0.00, 34.3)	16.0 (5.71, 34.5)	0.296
Sleep duration, h/day	6.1 ± 1.5	5.9 ± 1.4	6.3 ± 1.6	0.183
AHI, events/h	58.0 (26.0, 89.0)	52.0 (28.5, 87.0)	63.5 (21.0, 94.0)	0.989
Severe OSA, n (%) [d]	62 (74)	32 (76)	30 (71)	0.666
CPAP therapy, n (%)	69 (82)	36 (86)	33 (79)	0.393

Normally distributed and skewed continuous variables are presented as mean ± standard deviation and median (1st, 3rd quartile), respectively, while categorical variables are presented as absolute number (relative frequency). [a] Based on the chi-square test for categorical variables or the Student's t-test and the Mann–Whitney U test for normal and skewed continuous variables, respectively. p-values < 0.050 indicate statistically significant differences. [b] BMI $\geq$ 30 kg/m^2. [c] The score ranges from 0 to 55; higher values indicate higher adherence to the Mediterranean diet. [d] AHI $\geq$ 30 events/h of sleep. AHI, apnea-hypopnea index; BMI, body mass index; CPAP, continuous positive airway pressure; MedDietScore, Mediterranean diet score; MLG, Mediterranean lifestyle group; OSA, obstructive sleep apnea; SCG, standard care group.

At the 6-month follow-up, mean percent body weight change was 0.16 ± 2.91 % in the SCG and −11.1 ± 5.90 % in the MLG (p < 0.001). Patients in the SCG did not present any significant changes in lifestyle habits (data not shown); on the contrary, mean MedDietScore (from 31.7 ± 4.7 to 40.9 ± 3.9), median (1st, 3rd quartile) daily physical activity time [from 16.0 (5.71, 34.5) to 50.8 (42.9, 63.6) min/day] and mean sleep duration (from 6.3 ± 1.6 to 7.1 ± 0.57 h/day) increased in the MLG (all $p \leq 0.001$), with a significant difference compared to the SCG (all $p < 0.001$). According to the intention-to-treat analysis ($n = 84$), a decrease in median TNF-a levels was observed in the MLG ($p = 0.001$) but not in the SCG ($p = 0.975$) (Table 2). Median absolute and percent reduction of TNF-a was also higher in the MLG compared to the SCG (both $p < 0.050$). Post-intervention age-, sex-, baseline- and CPAP use-adjusted levels of TNF-a were lower in the MLG compared to the SCG [MD (95%CI): -0.97 (-1.74, -0.20) pg/mL, $p = 0.014$] and the difference between groups was not attenuated after further adjustment for %Δ weight ($p = 0.020$). When 6-month AHI was also included as a covariate in the model, the difference in TNF-a remained marginally significant ($p = 0.050$). Per protocol analyses ($n = 60$) revealed similar results (Table 2).

Table 2. Changes within groups and differences between groups in plasma TNF-a levels.

	Intention-to-Treat Analysis ($n = 84$)				
	SCG ($n = 42$)	p [a]	**MLG ($n = 42$)**	p [a]	p [b]
TNF-a BL (pg/mL)	2.98 (1.42, 4.09)	0.975	2.92 (2.37, 4.00)	0.001	0.986
TNF-a FU (pg/mL)	2.90 (2.58, 3.11)		2.00 (0.92, 3.30)		0.009
Δ TNF-a (pg/mL)	0.13 (-1.58, 1.78)		-0.73 (-2.24, 0.14)		0.029
%Δ TNF-a	11.1 (-36.3, 108)		-25.0 (-67.1, 6.85)		0.004
	Per Protocol Analysis ($n = 60$)				
	SCG ($n = 28$)	p [a]	**MLG ($n = 32$)**	p [a]	p [b]
TNF-a BL (pg/mL)	2.96 (1.30, 3.90)	0.685	2.97 (2.42, 4.18)	0.006	0.575
TNF-a FU (pg/mL)	2.90 (1.62, 3.60)		2.27 (0.79, 3.38)		0.283
Δ TNF-a (pg/mL)	0.63 (-1.29, 1.79)		-1.02 (-2.30, 0.28)		0.049
%Δ TNF-a	29.7 (-33.0, 117)		-29.5 (-70.0, 4.68)		0.015

Data are presented as median (1st, 3rd quartile). [a] Based on the Wilcoxon signed-rank test. p-values < 0.050 indicate statistically significant changes. [b] Based on the Mann–Whitney U test. p-values < 0.050 indicate statistically significant differences. BL, baseline; FU, follow-up; MLG, Mediterranean lifestyle group; SCG, standard care group; TNF-a, tumor necrosis factor-a; (%)Δ, (percent) change.

4. Discussion

In the present study, the effect of a weight-loss lifestyle intervention on plasma TNF-a levels was explored in an adult, overweight population of patients with OSA of at least moderate-severity. Although CPAP alone did not have a significant effect on TNF-a, a meaningful reduction of approximately 1 pg/mL was achieved when CPAP was combined with a feasible behavioral intervention aiming at a healthier body weight through the adoption of the Mediterranean lifestyle. This is in line with previous clinical trials showing improvements in TNF-a after healthy lifestyle interventions in individuals of increased cardiometabolic risk (e.g., those with metabolic syndrome) and patients with diabetes mellitus type 2 and cardiovascular disease, conditions which are tightly linked to OSA from a pathophysiological point of view [20,21]. Interestingly, the observed reduction in TNF-a was independent of weight loss and can be partly attributed to the strong anti-inflammatory properties of a healthy lifestyle, combining a Mediterranean-style diet [22] with daily physical activity [23] and adequate sleep duration [24].

OSA is tightly and bidirectionally linked to systemic inflammation [1]. Although several inflammatory markers have been positively associated with the presence of OSA, TNF-a in particular has emerged as a clinically-useful index for predicting OSA risk, with patients exhibiting higher TNF-a levels compared to healthy controls, and TNF-a values increasing as OSA severity progresses from mild to severe [5]. Given that TNF-a is implicated in cardiovascular pathophysiology [25], interventions for normalizing TNF-a values are important in OSA, which is currently recognized as a disease of cardiometabolic nature [26]. In this context, although CPAP remains the gold-standard therapy for OSA, its effectiveness in ameliorating inflammation remains controversial; some previous clinical trials have revealed that CPAP therapy can lead to significant reductions in markers of inflammation (e.g., CRP, TNF-a and interleukin-6) in patients with OSA [27], whereas other studies, similarly to our findings, have not reported a significant anti-inflammatory effect [28]. The observed lack of a significant effect of CPAP on TNF-a is in line with theories proposing that OSA represents a manifestation of the metabolic syndrome [29,30]. In this context, it can be speculated that chronic subclinical inflammation, a typical component of the metabolic syndrome, pre-exists and contributes to the onset and progression of OSA, combined with central obesity, insulin resistance and oxidative stress. Although the cause–effect link between inflammation and OSA requires further investigation, it is possible that the inflammatory state in OSA is not merely a consequence of sleep-disordered breathing and intermittent hypoxia, and therefore cannot be entirely reversed through CPAP, which aims at normalizing breathing during sleep. This is also supported by the fact that the difference between groups in TNF-a observed in our study was partly attenuated but remained marginally significant after adjustment for residual AHI; this observation suggests that improvements in OSA severity can only partially explain improvements in inflammation and that healthy lifestyle interventions can both improve OSA severity and ameliorate inflammation through other mechanisms, including beneficial effects on body weight status and the pathophysiology of the metabolic syndrome [31]. The suboptimal use of CPAP by study participants, which is in line with previously published data [32], is another possible explanation for the lack of a significant change in TNF-a levels in the SCG.

To our best knowledge, this is the first interventional study to explore the effects of a weight-loss lifestyle intervention on plasma TNF-a levels in OSA. The study design (randomized controlled clinical trial); the diagnosis of OSA through an attended overnight in-hospital PSG; the evaluation of patients' inflammatory status through TNF-a, which is considered a key modulator of systemic inflammation and an important biomarker for the onset and progression of OSA; and the implementation of a well-designed, feasible, multi-component behavioral lifestyle intervention based on the health-promoting Mediterranean pattern are strong points of the present work. Limitations of our study include: the fact that this was a secondary analysis in a subsample of the original study population with available TNF-a measurements; the small sample size ($n = 84$) and the relatively high drop-out rate ($n = 24$, 28.5% of the study population), although we partially compensated for attrition by implementing intention-to-treat analysis and achieved sufficient statistical power for analyses; the fact that CPAP was prescribed and not provided to patients, which could explain its suboptimal use (82% of the study population, mean use: 3.89 ± 2.47 h/day), although the parentage of CPAP users was similar between study groups and the daily duration of CPAP use was included as a covariate in analyses; and the fact that our study was implemented in Greece, a typical Mediterranean country, in which the Mediterranean diet could be more easy to adopt, well-accepted by patients and sustainable and therefore our results cannot be universally generalized in the whole OSA population, especially patients in non-Mediterranean countries.

In conclusion, our findings support that healthy lifestyle interventions can be an efficient approach for improving OSA-related systemic inflammation and should be further tested and confirmed in future clinical trials incorporating adequate samples of OSA patients of different ethnic, sociodemographic and lifestyle backgrounds.

Author Contributions: Conceptualization, N.Y. and M.D.K.; methodology, N.Y., E.V. and M.D.K.; formal analysis, M.G.; investigation, M.G., I.K., R.T. and K.L.; resources, N.Y. and E.V.; writing—original draft preparation, M.G.; writing—review and editing, N.Y., I.K., R.T., K.L., E.V. and M.D.K.; supervision, M.D.K.; project administration, N.Y. and M.D.K. All authors have read and agreed to the published version of the manuscript.

Funding: The study was partially funded by the Postgraduate Program "Applied Nutrition and Dietetics", Department of Nutrition and Dietetics, Harokopio University.

Institutional Review Board Statement: The study was conducted in accordance with the Declaration of Helsinki and approved by the Ethics Committee of HAROKOPIO UNIVERSITY (protocol code 46/3-4-2015 and date of approval 3 April 2015).

Informed Consent Statement: Informed consent was obtained from all subjects involved in the study.

Data Availability Statement: Deidentified participant data and the study protocol will be made available by the corresponding author upon request.

Acknowledgments: The authors are grateful to P. Ouzounidou, Center of Sleep Disorders, Evangelismos Hospital, Athens, Greece, and Antigoni Tsiafitsa, Department of Nutrition & Dietetics, Harokopio University, Athens, Greece for collecting patients' blood samples.

Conflicts of Interest: The authors declare no conflict of interest.

References

1. Unnikrishnan, D.; Jun, J.; Polotsky, V. Inflammation in sleep apnea: An update. *Rev. Endocr. Metab. Disord.* **2015**, *16*, 25–34. [CrossRef]
2. Galland, L. Diet and inflammation. *Nutr. Clin. Pract.* **2010**, *25*, 634–640. [CrossRef]
3. Barnes, M.; Goldsworthy, U.R.; Cary, B.A.; Hill, C.J. A diet and exercise program to improve clinical outcomes in patients with obstructive sleep apnea–a feasibility study. *J. Clin. Sleep Med.* **2009**, *5*, 409–415. [CrossRef]
4. Chirinos, J.A.; Gurubhagavatula, I.; Teff, K.; Rader, D.J.; Wadden, T.A.; Townsend, R.; Foster, G.D.; Maislin, G.; Saif, H.; Broderick, P.; et al. CPAP, weight loss, or both for obstructive sleep apnea. *N. Engl. J. Med.* **2014**, *370*, 2265–2275. [CrossRef]
5. Li, Q.; Zheng, X. Tumor necrosis factor alpha is a promising circulating biomarker for the development of obstructive sleep apnea syndrome: A meta-analysis. *Oncotarget* **2017**, *8*, 27616–27626. [CrossRef]
6. Georgoulis, M.; Yiannakouris, N.; Kechribari, I.; Lamprou, K.; Perraki, E.; Vagiakis, E.; Kontogianni, M.D. The effectiveness of a weight-loss Mediterranean diet/lifestyle intervention in the management of obstructive sleep apnea: Results of the "MIMOSA" randomized clinical trial. *Clin. Nutr.* **2021**, *40*, 850–859. [CrossRef]
7. Georgoulis, M.; Yiannakouris, N.; Tenta, R.; Fragopoulou, E.; Kechribari, I.; Lamprou, K.; Perraki, E.; Vagiakis, E.; Kontogianni, M.D. A weight-loss Mediterranean diet/lifestyle intervention ameliorates inflammation and oxidative stress in patients with obstructive sleep apnea: Results of the "MIMOSA" randomized clinical trial. *Eur. J. Nutr.* **2021**, *60*, 3799–3810. [CrossRef]
8. Georgoulis, M.; Yiannakouris, N.; Kechribari, I.; Lamprou, K.; Perraki, E.; Vagiakis, E.; Kontogianni, M.D. Cardiometabolic Benefits of a Weight-Loss Mediterranean Diet/Lifestyle Intervention in Patients with Obstructive Sleep Apnea: The "MIMOSA" Randomized Clinical Trial. *Nutrients* **2020**, *12*, 1570. [CrossRef]
9. Qaseem, A.; Holty, J.E.; Owens, D.K.; Dallas, P.; Starkey, M.; Shekelle, P.; Clinical Guidelines Committee of the American College of Physicians. Management of obstructive sleep apnea in adults: A clinical practice guideline from the American College of Physicians. *Ann. Intern. Med.* **2013**, *159*, 471–483. [CrossRef]
10. Patil, S.P.; Ayappa, I.A.; Caples, S.M.; Kimoff, R.J.; Patel, S.R.; Harrod, C.G. Treatment of Adult Obstructive Sleep Apnea with Positive Airway Pressure: An American Academy of Sleep Medicine Clinical Practice Guideline. *J. Clin. Sleep Med.* **2019**, *15*, 335–343. [CrossRef]
11. Bach-Faig, A.; Berry, E.M.; Lairon, D.; Reguant, J.; Trichopoulou, A.; Dernini, S.; Medina, F.X.; Battino, M.; Belahsen, R.; Miranda, G.; et al. Mediterranean diet pyramid today. Science and cultural updates. *Public Health Nutr.* **2011**, *14*, 2274–2284. [CrossRef]
12. *Global Recommendations on Physical Activity for Health*; WHO: Geneva, Switzerland, 2010.
13. Watson, N.F.; Badr, M.S.; Belenky, G.; Bliwise, D.L.; Buxton, O.M.; Buysse, D.; Dinges, D.F.; Gangwisch, J.; Grandner, M.A.; Kushida, C.; et al. Recommended Amount of Sleep for a Healthy Adult: A Joint Consensus Statement of the American Academy of Sleep Medicine and Sleep Research Society. *Sleep* **2015**, *38*, 843–844. [CrossRef]
14. Wadden, T.A.; Webb, V.L.; Moran, C.H.; Bailer, B.A. Lifestyle modification for obesity: New developments in diet, physical activity, and behavior therapy. *Circulation* **2012**, *125*, 1157–1170. [CrossRef]
15. Wadden, T.A.; Butryn, M.L.; Wilson, C. Lifestyle modification for the management of obesity. *Gastroenterology* **2007**, *132*, 2226–2238. [CrossRef]
16. Bountziouka, V.; Bathrellou, E.; Giotopoulou, A.; Katsagoni, C.; Bonou, M.; Vallianou, N.; Barbetseas, J.; Avgerinos, P.C.; Panagiotakos, D.B. Development, repeatability and validity regarding energy and macronutrient intake of a semi-quantitative food frequency questionnaire: Methodological considerations. *Nutr. Metab. Cardiovasc. Dis. NMCD* **2012**, *22*, 659–667. [CrossRef]
17. Panagiotakos, D.B.; Pitsavos, C.; Arvaniti, F.; Stefanadis, C. Adherence to the Mediterranean food pattern predicts the prevalence of hypertension, hypercholesterolemia, diabetes and obesity, among healthy adults; the accuracy of the MedDietScore. *Prev. Med.* **2007**, *44*, 335–340. [CrossRef]
18. Papathanasiou, G.; Georgoudis, G.; Georgakopoulos, D.; Katsouras, C.; Kalfakakou, V.; Evangelou, A. Criterion-related validity of the short International Physical Activity Questionnaire against exercise capacity in young adults. *Eur. J. Cardiovasc. Prev. Rehabil.* **2010**, *17*, 380–386. [CrossRef]

19. Little, R.J.; D'Agostino, R.; Cohen, M.L.; Dickersin, K.; Emerson, S.S.; Farrar, J.T.; Frangakis, C.; Hogan, J.W.; Molenberghs, G.; Murphy, S.A.; et al. The prevention and treatment of missing data in clinical trials. *N. Engl. J. Med.* **2012**, *367*, 1355–1360. [CrossRef]
20. Beavers, K.M.; Nicklas, B.J. Effects of lifestyle interventions on inflammatory markers in the metabolic syndrome. *Front. Biosci.* **2011**, *3*, 168–177. [CrossRef]
21. Gaesser, G.A.; Angadi, S.S.; Ryan, D.M.; Johnston, C.S. Lifestyle Measures to Reduce Inflammation. *Am. J. Lifestyle Med.* **2012**, *6*, 4–13. [CrossRef]
22. Casas, R.; Sacanella, E.; Estruch, R. The immune protective effect of the Mediterranean diet against chronic low-grade inflammatory diseases. *Endocr. Metab. Immune Disord. Drug Targets* **2014**, *14*, 245–254. [CrossRef]
23. Lin, X.; Zhang, X.; Guo, J.; Roberts, C.K.; McKenzie, S.; Wu, W.C.; Liu, S.; Song, Y. Effects of Exercise Training on Cardiorespiratory Fitness and Biomarkers of Cardiometabolic Health: A Systematic Review and Meta-Analysis of Randomized Controlled Trials. *J. Am. Heart Assoc.* **2015**, *4*, e002014. [CrossRef]
24. Irwin, M.R. Sleep and inflammation: Partners in sickness and in health. *Nat. Rev. Immunol.* **2019**, *19*, 702–715. [CrossRef]
25. Rolski, F.; Blyszczuk, P. Complexity of TNF-alpha Signaling in Heart Disease. *J. Clin. Med.* **2020**, *9*, 3267. [CrossRef]
26. Floras, J.S. Sleep Apnea and Cardiovascular Disease: An Enigmatic Risk Factor. *Circ. Res.* **2018**, *122*, 1741–1764. [CrossRef]
27. Xie, X.; Pan, L.; Ren, D.; Du, C.; Guo, Y. Effects of continuous positive airway pressure therapy on systemic inflammation in obstructive sleep apnea: A meta-analysis. *Sleep Med.* **2013**, *14*, 1139–1150. [CrossRef]
28. Ning, Y.; Zhang, T.S.; Wen, W.W.; Li, K.; Yang, Y.X.; Qin, Y.W.; Zhang, H.N.; Du, Y.H.; Li, L.Y.; Yang, S.; et al. Effects of continuous positive airway pressure on cardiovascular biomarkers in patients with obstructive sleep apnea: A meta-analysis of randomized controlled trials. *Sleep Breath. Schlaf Atm.* **2019**, *23*, 77–86. [CrossRef]
29. Vgontzas, A.N.; Gaines, J.; Ryan, S.; McNicholas, W.T. CrossTalk proposal: Metabolic syndrome causes sleep apnoea. *J. Physiol.* **2016**, *594*, 4687–4690. [CrossRef]
30. Vgontzas, A.N.; Bixler, E.O.; Chrousos, G.P. Sleep apnea is a manifestation of the metabolic syndrome. *Sleep Med. Rev.* **2005**, *9*, 211–224. [CrossRef]
31. van Namen, M.; Prendergast, L.; Peiris, C. Supervised lifestyle intervention for people with metabolic syndrome improves outcomes and reduces individual risk factors of metabolic syndrome: A systematic review and meta-analysis. *Metab. Clin. Exp.* **2019**, *101*, 153988. [CrossRef]
32. Askland, K.; Wright, L.; Wozniak, D.R.; Emmanuel, T.; Caston, J.; Smith, I. Educational, supportive and behavioural interventions to improve usage of continuous positive airway pressure machines in adults with obstructive sleep apnoea. *Cochrane Database Syst. Rev.* **2020**, *4*, CD007736. [CrossRef]

Article

The Changes in the Severity of Deep Neck Infection Post-UPPP and Tonsillectomy in Patients with OSAS

Pin-Ching Hu [1,†], Liang-Chun Shih [1,2,3,†], Wen-Dien Chang [4], Jung-Nien Lai [5], Pei-Shao Liao [1], Chih-Jaan Tai [1,3,6,7], Chia-Der Lin [1,3,6], Hei-Tung Yip [7,8], Te-Chun Shen [6,9] and Yung-An Tsou [1,2,3,6,*]

[1] Department of Otolaryngology-Head and Neck Surgery, China Medical University Hospital, Taichung 404332, Taiwan
[2] Department of Otolaryngology-Head and Neck Surgery, Asia University Hospital, Taichung 40002, Taiwan
[3] Graduate Institute of Biomedical Sciences, China Medical University, Taichung 40402, Taiwan
[4] Department of Sport Performance, National Taiwan University of Sport, Taichung 404401, Taiwan
[5] School of Chinese Medicine, College of Chinese Medicine, China Medical University, Taichung 40402, Taiwan
[6] School of Medicine, China Medical University, Taichung 40402, Taiwan
[7] Department of Health Services Administration, China Medical University, Taichung 40402, Taiwan
[8] Management Office for Health Data, China Medical University Hospital, Taichung 404332, Taiwan
[9] Division of Pulmonary and Critical Care Medicine, Department of Internal Medicine, China Medical University Hospital, Taichung 40447, Taiwan
* Correspondence: d22052121@gmail.com
† These authors contributed equally to this work.

Citation: Hu, P.-C.; Shih, L.-C.; Chang, W.-D.; Lai, J.-N.; Liao, P.-S.; Tai, C.-J.; Lin, C.-D.; Yip, H.-T.; Shen, T.-C.; Tsou, Y.-A. The Changes in the Severity of Deep Neck Infection Post-UPPP and Tonsillectomy in Patients with OSAS. *Life* **2022**, *12*, 1196. https://doi.org/10.3390/life12081196

Academic Editor: Konstantinos Chaidas

Received: 26 June 2022
Accepted: 3 August 2022
Published: 5 August 2022

Publisher's Note: MDPI stays neutral with regard to jurisdictional claims in published maps and institutional affiliations.

Abstract: The main aim of this study is to compare the incidence rate and severity of deep neck infection (DNI) in patients post-UPPP+ T (uvulopalatopharyngoplasty plus tonsillectomy) and without UPPP+ T. We utilized the data derived from the Longitudinal Health Insurance Database (LHID) of the National Health Insurance Research Database (NHIRD) in Taiwan from 1 January 2000 to 31 December 2012. Patients who had undergone combined UPPP and tonsillectomy were selected using National Health Insurance (NHI) surgical order. Patients with DNI were selected using International Classification of Diseases (ICD-9-CM) code. A logistic regression model was applied for risk analysis. There were 1574 patients in the UPPP+ T cohort, and 6,296 patients who did not undergo combined UPPP and tonsillectomy for the control group. Our analysis showed that patients with an obstructive sleep apnea syndrome (OSAS) history constitute 76.1% (n = 1198) of the UPPP+ T cohort. Compared to the control group, there was no significantly increased incidence rate of DNI after UPPP+ T within 1–60 months. Patients undergoing combined UPPP and tonsillectomy had a lower intubation rate for DNI, with an adjusted odds ratio of 0.47 (95% CI = 0.32–0.69). The combined UPPP and tonsillectomy does not increase the risk of DNI within 1–60 months. Furthermore, combined UPPP and tonsillectomy can reduce the severity for DNI by decreasing the intubation rate and length of hospitalization.

Keywords: deep neck infection; uvulopalatopharyngoplasty; obstructive sleep apnea syndrome; National Health Insurance Research Database

1. Introduction

Deep neck infection (DNI) is a severe but treatable infection of deep cervical spaces, featuring rapid disease progression and many life-threatening complications [1]. Potential causes of DNI include odontogenic infections, salivary origin infections, pharyngitis, tonsillitis, cervical lymphadenitis, and trauma to the head and neck. There were several deep cervical spaces frequently related to tonsillitis, including peritonsillar space, submandibular space, parapharyngeal spaces, carotid spaces, and retropharyngeal space. Tonsillitis related peritonsillar abscess and pharyngeal spaces (parapharyngeal spaces, carotid spaces, retropharyngeal space) are considerably correlated. Quisy tonsillectomy is also considered as a therapy for peritonsillar abscess and repeated tonsillitis. Klebsiella pneumoniae,

Staphylococcus aureus, and streptococci are common aerobic pathogens; Bacteroides and peptostreptococcus were the common nonaerobic pathogen. Surgical incision and drainage of deep neck abscess is warranted if empiric antibiotics cannot control the infectious condition. Disease progression often leads to sepsis and compromised airway condition. Diabetes mellitus, HIV infection, intravenous drug abuse, steroid therapy, chemotherapy, and other immune dysfunction diseases are common risk factors of DNI [1,2].

Obstructive sleep apnea syndrome (OSAS) is a sleep disorder featuring repeated episodes of apnea or hypopnea during sleep and intermittent arousals from sleep, which is a result of total and/or partial collapse of upper airways [3]. The disruption of breathing leads to intermittent hypoxemia and sympathetic activation [4,5], oxidative stress, systemic inflammation, and metabolic changes, resulting in cardiovascular and metabolic morbidities [5]. Current research showed that OSAS associated with known cardiovascular risk factors, such as obesity, insulin resistance, and dyslipidemia [6]. Antonopoulou et al. indicated that local and systemic inflammation relate to the pathophysiology of OSAS [7]. Inancli et al. also have found that OSAS associates with upper airway inflammation, and the inflammatory processes may be potential mediators of cardiovascular morbidity in these patients who have OSAS [8]. OSAS affects approximately 24% of men and 9% of women [9]. Obesity, male sex, and aging are major risk factors of OSAS [4].

Upper airway anatomy, such as airway length, lateral pharyngeal wall thickness, tongue volume and dilator muscle activity are important mechanisms that affect the occurrence of OSAS [4,10]. When it comes to surgical therapy for OSAS, uvulopalatopharyngoplasty (UPPP) combined with tonsillectomy is a common and widely performed surgery, which typically involves resection of the uvula, palatine tonsils, and posterior segment of the soft palate, remodeling the retropalatal airway and decreasing the collapsibility of the upper airway [11].

As stated before, many studies have addressed UPPP+ T (uvulopalatopharyngoplasty plus tonsillectomy) and its surgical response in the cardiovascular system [12,13]. Long-term complications of UPPP+ T include velopharyngeal insufficiency, pharyngeal symptoms, such as tightness or dryness, taste/voice disturbance, and nasopharyngeal stenosis [11]. However, the relationship between UPPP+ T and local infection, especially DNI, remains unclear. In our research, we compared the risk and severity of DNI in the patients with post-UPPP+ T to those without UPPP+ T.

2. Materials and Methods

2.1. Data Source

In this cohort study, we used the data from the Taiwan National Health Insurance Research Database (NHIRD), which was established by the Taiwan government in 1995. The study period is from 2000 to 2012. The treatment combined UPPP and tonsillectomy is all performed between 2000 to 2007, the occurrence of DNI is observed after operation or till to end of 2012. We utilized the data derived from the Longitudinal Health Insurance Database (LHID) of the NHIRD in Taiwan from 1 January 2000, to 31 December 2012. This is because of the NHI authority only allowed us to analyze that time span and thus we could only survey that time period. Almost all of Taiwan's residents participated in the National Health Insurance program and the health information is stored in NHIRD. The information in the database includes medical services, devices, and prescription drug records. The diagnosis code recorded in the database followed the International Classification of Diseases, Ninth Revision, Clinical Modification (ICD-9-CM) code. The subset data, LHID, containing data of one million randomly selected insurances from the NHIRD, was the main data source. P.-C.H, L.-C.S., and Y.-A.T. collected the data and L.-C.S., Y.-A.T., C.-J.T., and C.-D.L. checked each datum and discussed the real DNI condition that was defined as an infection event. W.-D.C., J.-N.L., P.-S.L., H.-T.Y., and T.-C.S. rechecked the collected data and conducted the data analyses. This study was approved by the Institutional Review Board of China Medical University Hospital.

2.2. Study Population

Our study was a case cohort study, considering patients who underwent UPPP+ T between 2000 and 2012. Those without UPPP+ T were the control patients. Four control patients were matched to a case patient by sex, age, index year, and comorbidities using propensity score matching. The index date of the case group was the date of received UPPP+ T and that of the control group was a random date within the study period. We eliminated patients developed outcomes of interest before the index date, with a follow-up time less than six months and all aged below 18. All subjects were followed until the occurrence of primary outcome, withdraw from the program, or the end of 2013.

2.3. Main Outcome and Comorbidities

Deep neck infection (DNI) (ICD-9-CM code 475, 478.22, 478.24 527.3, 528.3, 682.0, 682.1) was the primary outcome of this study. We excluded DNI that occurred within a month after UPPP+ T because physicians often prescribe post-operation antibiotics to prevent tonsillitis or DNI. Patients who had undergone UPPP+ T were selected using national health insurance (NHI) surgical order (66025B). The UPPP+ T is not only for treating sleep apnea or snoring, but also for repeated tonsillitis and chronic pharyngeal inflammation [14]. So, as for UPPP+ T subgroup analyzation, after excluding deviated nasal septum disease (ICD-9-CM code 470), other diseases of upper respiratory tract (ICD-9-CM code 478) and other sleep disturbances (ICD-9-CM code 780.59), we further divided UPPP+ T cohort into OSAS-caused group and inflammation/infection group. Obstructive sleep apnea syndrome (OSAS) was defined by ICD-9-CM code 780.51, 780.53, 780.57, 327.20, 327.23, 327.29, 327.8, 780.50, 786.09. Inflammation and infection group was defined by ICD-9-CM code 474.0, A315, 474.0, A319, 472.0, 463, 474.11. The procedures of UPPP+ T included tonsillectomy and pharyngoplasty. One of the indications of UPPP+ T is chronic tonsillitis with snoring [15]. Chronic tonsillitis means 3–4 occurrences of tonsillitis per year [16]. The palatine tonsils represent the nidus of inflammation, and therefore surgery, such as combined UPPP and tonsillectomy, involving the removal of infection source in the tonsil, could treat the tonsillar-related infection. UPPP+ T is not only used to treat chronic tonsillitis, but also breathing disorders like snoring. To clarify whether the risk of DNI increased after UPPP+ T, we analyzed the incidence rate of DNI after 1–60 months between the control group and UPPP+ T cohort.

For further study, we also looked into the length of hospital stay, emergency department (ED) admission and stayed in intensive care unit (ICU) to determine the seriousness of DNI [17]. Airway complications such as intubation (NHI procedure code 47031C), tracheostomy (NHI procedure code 56022C) and oxygen inhalation (NHI procedure code 57003C and 57004C) were observed to assess the progression and severity for DNI. Diabetes mellitus, hypercholesterolemia, overweight and obesity, depression, hypertension, deviated nasal septum, nasal polyps, hypertrophy of tonsils and adenoids, asthma, and gastroesophageal reflux disease were considered as comorbidities of UPPP+ T.

2.4. Statistical Analysis

To examine the distribution of sex, age group, and comorbidities between the case cohort and the control cohort, the chi-square test was used. The average age among two groups was tested by the Student's t-test. Univariable and multivariable Cox proportional hazard models were employed to estimate the crude hazard ratio (cHR) and the adjusted hazard ratio (aHR). The Kaplan–Meier method was applied to obtain the cumulative incidence curve and testing was performed using the log-rank test. Odds ratios were estimated by the logistic regression model, and the relationship of changes was analyzed using simple linear regression. All statistical analyses were conducted using R Statistical Software, version 3.5.2 and SAS software, version 9.4 (SAS Institute Inc., Cary, NC, USA).

3. Results

A total of 1574 patients with UPPP+ T and 6296 patients without UPPP+ T were recruited in our cohort study. Refer to Table 1, male patients were dominant in the participants. Most of them were in the 30–39 year-old age group and the mean age of the controls and cases were 40.3 and 38.7. The UPPP+ T patients had a higher proportion of hypertrophy of tonsils and adenoids than non-UPPP+ T patients.

Table 1. Baseline characteristics of patients.

	Non-UPPP+ T No (n = 6296)		UPPP+ T Yes (n = 1574)		
Variable	n	%	n	%	p-**Value**
Gender					0.76
female	1636	26.0	415	26.4	
male	4660	74.0	1159	73.6	
Age, year					0.87
18–29	1695	26.9	425	27.0	
30–39	1826	29.0	463	29.4	
40–49	1545	24.5	393	25.0	
$\geq$50	1230	19.5	293	18.6	
mean, (SD)	40.26	(14.9)	38.74	(12.3)	<0.001
Comorbidities					
diabetes mellitus	700	11.1	180	11.4	0.72
hypercholesterolemia	692	11.0	168	10.7	0.72
overweight and obesity	189	3.0	46	2.9	0.87
depression	650	10.3	148	9.4	0.28
hypertension	1428	22.7	349	22.2	0.67
deviated nasal septum	1513	24.0	352	22.4	0.16
nasal polyps	96	1.5	27	1.7	0.59
hypertrophy of tonsils and adenoids	169	2.7	66	4.2	0.002
asthma	880	14.0	219	13.9	0.95
gastroesophageal reflux disease	141	2.2	35	2.2	0.97

n: number of patients; UPPP: uvulopalatopharyngoplasty; SD: standard deviation.

Table 2 demonstrates the incidence and hazard ratio of baseline factors for DNI. Relative to those subjects aged 18–29, the DNI risk is higher in 30–39, 40–49, and over 50 age group, especially subjects aged 40–49, where the risk of DNI increased by 1.52 folds (95% CI = 1.01–2.30). Patients with diabetes mellitus will have an increased risk of DNI (aHR = 1.73; 95% CI = 1.12–2.67).

In Tables 3 and 4, we perform a subgroup analysis, which showed that patients with OSAS history constitute 76.1% (n = 1198) of the UPPP+ T cohort. Compared to the control group, there was no significantly increased incidence rate of DNI in UPPP+ T cohort within 1–60 months (p > 0.05), as shown in Table 3. The slope of the regression line for the incidence rate of DNI for UPPP+ T (r = −0.92, R^2 = 0.45) and non-UPPP+ T (r = 0.08, R^2 = 0.06) decreased over follow-up, indicating that incidence rate of DNI decreased over the course for UPPP+ T. The difference in the slope change of the incidence rate of DNI was statistically significant (p < 0.05). The incidence rate and hazard ratios of DNI between UPPP+ T subgroups have no significant difference (Table 4).

The mean hospital stay of the UPPP+ T cohort (5.20 days) was significantly shorter than non-UPPP+ T cohort (7.81 days), with an adjusted relative ratio of 0.60, and patients in the inflammation/infection group were not hospitalized (Table 5). In the non-UPPP+ T cohort, 4.41% patients (n = 6) were admitted from the emergency department (ED), 0.74% patients (n = 1) were transferred to the intensive care unit (ICU), while the UPPP+ T cohort had only 1.59% patients (n = 1) admitted from ED and no case transferred to ICU (Table 5). Considering non-UPPP+ T patients as the reference group, the UPPP+ T cohort had a lower intubation rate, with an adjusted odds ratio of 0.47 (95% CI = 0.32–0.69) (Table 6).

Table 2. Incidence and hazard ratio of baseline factors for Deep Neck Infection.

			Deep Neck Infection				
Variables	n	PY	IR	cHR	(95% CI)	aHR ‡	(95% CI)
Gender							
Female	61	15,216	4.01	1.00	(reference)	1.00	(reference)
Male	138	39,428	3.50	0.87	(0.64, 1.18)	0.85	(0.63, 1.15)
Age, year							
18–29	42	15,664	2.68	1.00	(reference)	1.00	(reference)
30–39	63	16,342	3.86	1.45	(0.98, 2.15)	1.44	(0.97, 2.14)
40–49	58	13,876	4.18	1.58	(1.06, 2.35) *	1.52	(1.01, 2.30) *
≥50	36	8762	4.11	1.51	(0.97, 2.37)	1.38	(0.83, 2.27)
Comorbidities							
Diabetes mellitus							
No	170	50,049	3.40	1.00	(reference)	1.00	(reference)
Yes	29	4595	6.31	1.80	(1.21, 2.68) **	1.73	(1.12, 2.67) *
Hypercholesterolemia							
No	186	50,327	3.70	1.00	(reference)	1.00	(reference)
Yes	13	4317	3.01	0.78	(0.45, 1.38)	0.65	(0.36, 1.17)
Overweight and obesity							
No	194	53,589	3.62	1.00	(reference)	1.00	(reference)
Yes	5	1055	4.74	1.25	(0.51, 3.04)	1.05	(0.42, 2.6)
Depression							
No	180	50,440	3.57	1.00	(reference)	1.00	(reference)
Yes	19	4204	4.52	1.23	(0.76, 1.97)	1.15	(0.71, 1.85)
Hypertension							
No	157	45,060	3.48	1.00	(reference)	1.00	(reference)
Yes	42	9584	4.38	1.22	(0.86, 1.71)	1.08	(0.73, 1.6)
Deviated nasal septum							
No	165	44,173	3.74	1.00	(reference)	1.00	(reference)
Yes	34	10,471	3.25	0.84	(0.58, 1.22)	0.86	(0.59, 1.25)
Nasal polyps							
No	198	54,108	3.66	1.00	(reference)	1.00	(reference)
Yes	1	536	1.86	0.49	(0.07, 3.5)	0.54	(0.08, 3.94)
Hypertrophy of tonsils and adenoids							
No	193	53,440	3.61	1.00	(reference)	1.00	(reference)
Yes	6	1204	4.98	1.33	(0.59, 3.00)	1.31	(0.58, 2.98)
Asthma							
No	180	48,701	3.70	1.00	(reference)	1.00	(reference)
Yes	19	5943	3.20	0.84	(0.52, 1.34)	0.84	(0.52, 1.35)
Gastroesophageal reflux disease							
No	197	54,174	3.64	1.00	(reference)	1.00	(reference)
Yes	2	470	4.26	1.11	(0.27, 4.49)	1.04	(0.25, 4.23)

n: number of patients; PY: person-year; IR: incidence rate pre 1000 person-years; cHR: crude hazard ratio; aHR: adjusted hazard ratio; CI: confidence interval; ‡: adjusted by sex, age and comorbidities; *: *p*-value < 0.05; **: *p*-value < 0.01.

Table 3. Analysis of the incidence rate of DNI among the follow-up.

				Incidence rate of DNI per 1000 Person-Years				
	n	Event [a]	Baseline	1–3 Months	3–6 Months	6–12 Months	12–36 Months	36–60 Months
Non-UPPP+ T	6296	136	2.16	2.56	3.96	3.46	3.44	3.23
UPPP+ T	1574	63	4.01	7.66	10.40	5.41	5.96	5.25
p value				0.15	0.23	0.54	0.18	0.46
OSAS cause	1198	41	3.42	10.07	10.27	5.39	4.48	6.61
p value				0.15	0.33	0.22	0.27	0.19
Inflammation and infection	195	7	3.58	0	0	0	1.92	0
p value				na	na	na	0.93	na

n: number of patients; DNI: deep neck infection; OSAS: Obstructive sleep apnea syndrome; UPPP: uvulopalatopharyngoplasty; na, not applicable; [a] number of DNI before the interventions; * *p* < 0.05, compare to baseline by using chi-square test.

Table 4. Incidences rate and hazard ratios of DNI between the UPPP+ T subgroups.

	n	Event [a]	PY	Incidence Rate	Crude HR	Adjusted HR
UPPP+ T						
OSAS-caused	1198	41	8219	4.99	1 (reference)	1 (reference)
Inflammation and infection	195	7	1811	3.87	0.81 (0.36, 1.81)	0.82 (0.36, 1.85)

n: number of patients; PY: person-year; Crude HR: crude hazard ratio; Adjusted HR: adjusted hazard ratio; [a] number of DNI before the interventions.

Table 5. S Severity of DNI between UPPP+ T and non-UPPP+ T groups.

	Hospital Stay		ED	ICU
	Mean, (SD)	aRR (95% CI)	*n* (%)	*n* (%)
Non-UPPP+ T	7.81 (5.50)	1 (reference)	6 (4.41%)	1 (0.74%)
UPPP+ T	5.20 (1.81)	0.60 (0.43, 0.84) **	1 (1.59%)	0 (0%)
OSAS cause	5.40 (0.98)	0.62 (0.40, 0.96) *	0 (0%)	0 (0%)
Inflammation and infection	-	-	0 (0%)	0 (0%)

n: number of events; CI: confidence interval; SD: standard deviation; UPPP+ T: uvulopalatopharyngoplasty plus tonsillectomy; ED: emergency department admission; ICU: intensive care unit; DNI: deep neck infection; OSAS: Obstructive sleep apnea syndrome;.adjusted RR: adjusted by sex, age and comorbidities; *: p-value <0.05; **: p-value <0.01.

Table 6. Airway complication rate.

	Intubation		Tracheostomy		Oxygen Inhalation	
	n (%)	Adjusted OR (95% CI)	n (%)	Adjusted OR (95% CI)	*n* (%)	Adjusted OR (95% CI)
Non-UPPP+ T	251 (4.02%)	1.00 (reference)	31 (0.49%)	1.00 (reference)	720 (12.9%)	1.00 (reference)
UPPP+ T	31 (1.97%)	0.47 (0.32, 0.69) ***	13 (0.83%)	1.74 (0.90, 3.35)	188 (14.8%)	1.18 (0.98, 1.41)

n: number of events; OR: odds ratio; CI: confidence interval; UPPP+ T: uvulopalatopharyngoplasty plus tonsillectomy; adjusted OR: adjusted by sex, age and comorbidities; ***: p-value <0.001.

4. Discussion

The combined UPPP and tonsillectomy including tonsillectomy is indicated for sleep apnea and chronic tonsillitis with snoring. It is considered to cause an immune compromising condition due to the partial removal of immune tissue at the pharynx. However, the risk of deep neck infection is not significant after combined UPPP and tonsillectomy within 1–60 months. In addition, we found that combined UPPP and tonsillectomy not only reduces the severity for DNI by decreasing intubation rate and shortening the length of hospitalization. In comparison to non-UPPP+ T, the baseline of DNI incident rate was higher in UPPP+ T. In fact, the surgical procedure of combined UPPP and tonsillectomy includes tonsillectomy, uvulectomy, and suture palatoplasty. Therefore, the palatine tonsils are removed after combined UPPP and tonsillectomy. Thus, the possible tonsillitis is treated, decreasing the inflammatory condition caused by chronic tonsillitis. In addition, UPPP+ T is also adopted as a surgery for sleep apnea and intermittent hypoxic condition. It is possible that improved sleep apnea or less intermittent hypoxemia after UPPP+ T also rendered anti-inflammatory therapeutic results that reduced the incidence of deep neck infection. In general, the incidence rate is higher in the UPPP+ T group than non-UPPP+ T group. With time, there was a more prominent decrease in DNI incidence rate than in the non-UPPP+ T group, as shown in Figure 1.

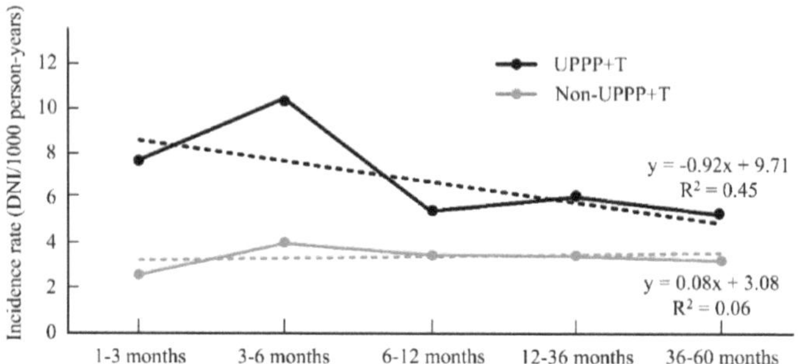

Figure 1. Changes in the slopes of DNI incidence rate.

4.1. Risk Factors for Deep Neck Infection

A review by Knapp et al. concluded that it is a common concept that diabetic mellitus (DM) patients can be infected more easily [18]. Moreover, it is known that DM is a common risk factor of DNI [1,2]. In line with previous studies, we found that the risk of DNI in the DM group was significantly higher than the non-DM group, with an adjusted hazard ratio of 1.73 ($p < 0.05$; Table 2). As stated before, odontogenic infection is the main cause of DNI. Adoviča et al. indicated that the occurrence rate of dental infection was significantly lower in elders because of the fewer teeth in their oral cavity [19]. Another study by Zamiri et al. also found that in the sixth, seventh, and eighth decades, the incidence rate of dental infection was lowest [20]. Our study had a similar result in that the adjusted hazard ratio was lower in the age group over 50 than age groups 40–49 and 30–39 (Table 2). However, contrary to the findings of previous research, our finding showed that compared to control group (age group 18–29), the DNI risk was higher in age group 30–39, 40–49, and over 50 (Table 2). Especially in age group 40–49, there was a significantly higher risk of DNI than the control group, with an adjusted hazard ratio of 1.52 ($p < 0.05$), as shown in Table 2. An apparent concern is that although the major infection source of DNI is dental infection, DNI presents less in elder patients, possible because they are more concerned about their teeth and have better hygiene. Besides, other infection sources exist, such as tonsillitis and pharyngitis. Furthermore, it is also important to consider the comorbidity in elder people. In the UPPP+ T group composing OSA that may also impact immunity, poor sleep quality leads to an immune compromising condition. There were also case of tonsillitis in the UPPP+ T group. In the combined UPPP and tonsillectomy, the tonsillectomy study illustrated that surgery has a protective effect in the case of cardiac problems, dementia, etc. [13,21,22] performed in Taiwan and Korea [23]. There were also studies showing the impact of adenotonsillectomy on the evolution of inflammatory markers that decreases tonsillitis related infection.

4.2. Risk of Deep Neck Infection after Uvulopalatopharyngoplasty

From our main result, we found that in patients with DNI, the patients without combined UPPP and tonsillectomy had a more consistent and lower incidence rate. However, compared to the control group, there was no significantly increased incidence rate of DNI after combined UPPP and tonsillectomy within 1–60 months (Table 3). This may imply that combined UPPP and tonsillectomy does not increase the risk of DNI within 1–60 months. Combined UPPP and tonsillectomy involves resection of the uvula, tonsils, and soft palate [11]. The tonsils provide the first line of protection against foreign pathogens, such as bacteria and viruses, but there is still a debate on whether tonsillectomy has a negative effect on the immune system [24]. We reviewed previous studies about the impact of immune system after tonsillectomy and two opposing opinions appeared (Table 7). Kaygusuz et al. found that long-term immune function shows no difference with healthy controls after tonsillectomy [25]. In a recent systematic review by Altwairqi et al., tonsillectomy had no negative affect on both humeral and cellular immunity in children and the level of immunoglobulin would recover to normal range postoperatively [26]. The other systematic review also showed that tonsillectomy has no negative clinical or immunological sequalae on the immune system [27]. A similar pattern of results was obtained in our study, showing that combined UPPP and tonsillectomy does not increase the risk of DNI within 1–60 months.

Table 7. Reviewed studies—impact of tonsillectomy on immune system.

Study	Year	Study Design	Main Results
Positive Opinion			
Kaygusuz et al. [25]	2009	cross-sectional	Tonsillectomy does not impair long-term (54 months) humoral and cellular immunity of children compared to their early-stage immune status (1 month). Moreover, the long-term (54 months) immune function has no different than healthy controls.
Bitar et al. [27]	2015	systematic review	Tonsillectomy has no negative clinical or immunological sequalae on the immune system.
Altwairqi et al. [26]	2020	systematic review	Tonsillectomy has no negative affect on both humeral and cellular immunity in children.
Negative Opinion			
Duval et al. [28]	2008	retrospective case-control study	Adenotonsillectomy would change the humoral and cellular response of the immune system in children.
Wang et al. [24]	2015	cohort study	Risk of DNI increased after tonsillectomy
Byars et al. [29]	2018	cohort study	Early-life (before age 9) tonsillectomy and adenoidectomy were associated with higher long-term (age 30) risks of respiratory, infectious, and allergic diseases.

In contrast, Wang et al. identified that the risk of DNI significantly increased among patients who have undergone tonsillectomy [24], and Duval et al. reported that adeno-tonsillectomy would change the humoral and cellular response of the immune system in children [28]. Another cohort study demonstrated that early-life tonsillectomy and adenoidectomy were associated with higher long-term risks of respiratory and infectious diseases [29]. However, very few publications mention the influence of tonsillectomy in adulthood. Whether tonsillectomy affects adult immune system still needs more investigation.

4.3. Adjustment of Obstructive Sleep Apnea Syndrome to Deep Neck Infection

Previous studies have indicated that the most common cause of DNI among adults is odontogenic or of salivary origin, such as dental and periodontal infection [1,2]. Normally, the oral biome is in balance, but mouth breathing in OSAS patients leads to dryness of the oral cavity and may also decrease the self-cleaning ability of the oral mucosa, leading to increased periodontal microbiota colonization and a chronic inflammatory response [30,31]. Xu et al. also revealed that changes of oxygen concentrations in the oral cavity in OSA patients might be associated with oral dysbiosis [32]. In a recent research by Ding et al. [33], there is a significantly higher occurrence of DNI in patients with sleep apnea. According to above research, there may be a correlation between DNI and OSAS.

In our study, we found that the incidence rate of DNI between UPPP+ T and non-UPPP+ T cohort had no significant difference (Table 3), even though a higher OSAS composition of combined UPPP and tonsillectomy subgroup (76.1%) might link to higher hypoxia-related inflammation and DNI risk. The combined UPPP and tonsillectomy is the main scope of our study. Moreover, the direction of selecting the research object and experimental design were different at the beginning, possibly explaining the different results compared to the above studies. However, a larger sample size study is still warranted in the future.

4.4. Severity of Deep Neck Infection after Uvulopalatopharyngoplasty

Based on our result, we found that combined UPPP and tonsillectomy can reduce the length of hospitalization and intubation rate of DNI. Length of hospitalization is often used as an indicator of infection severity. As reported by Sakarya et al., the average hospital stay of DNI patients was 12.9 days [34]. A single-center analysis by Kauffmann et al. showed that the mean duration of DNI hospital stay was 15.3 days, and patients with diabetes mellitus had a significantly longer duration of hospitalization [35]. Another retrospective review conducted by University of Kentucky focused on the postoperative length of stay in patients with DNI, which showed the overall hospitalization period was three days [36]. These studies also indicated that age, comorbidity, such as diabetes mellitus, development of complications, and treatment only with medicine service led to longer hospital stay days [34–36]. From our result, in patients with DNI, the mean hospital stay of the UPPP+ T cohort (5.20 days) was significantly lower than the non-UPPP+ T cohort (7.81 days) (Table 5). Furthermore, in the non-UPPP+ T cohort, 4.41% patients ($n = 6$) were admitted from the emergency department (ED) and 0.74% patients ($n = 1$) transferred to intensive care unit (ICU), while the UPPP+ T cohort had only 1.59% patients ($n = 1$) admitted from ED and no case transferred to ICU (Table 5). This may imply that combined UPPP and tonsillectomy can reduce the severity of DNI. However, it must be pointed out that the duration of hospitalization of DNI in our study was shorter than most other research [34,35]. In Taiwan, it is very convenient for the public to seek medical treatment because of the implementation of Taiwan's health insurance system. Hence, people in Taiwan get medical intervention earlier before the diseases progress.

We also observed airway complications, such as intubation, tracheostomy, and oxygen inhalation, to analyze the severity of DNI. Compared to the control group, the UPPP+ T cohort had a lower intubation rate (aOR = 0.47, $p < 0.001$; Table 6). Tracheostomy rate and oxygen using the rate of UPPP+ T cohort showed no significant difference compared to control group. Based on our result, it may imply that combined UPPP and tonsillectomy can reduce the intubation rate in DNI patients.

4.5. Limitations

Our study had some limitations. First, this study used the data from the Taiwan National Health Insurance Database (NHIRD), which may be affected by regional characteristics and cannot fully represent the overall situation of the world. Moreover, Taiwan's national health insurance allows people to get medical treatment regardless of their socio-economic status. Hence, some data may not be generalized to other populations. Second, the crucial data for surgical decision making, such as the apnea-hypopnea index and body mass index, are not available in our study, since NHIRD does not include biometric data of patients. Third, the incidence of multilevel combined UPPP and tonsillectomy increased from 2000 to 2012 in Taiwan [37]. Since NHIRD does not record patients' medical records, and whether combined UPPP and tonsillectomy is combined with other palatal, nasal, hypopharyngeal, or tongue surgery, it may affect the subsequent incidence of DNI and complication rate. Fourth, length of hospitalization as a severity indicator of DNI, may vary from institution to institution because of different medical resources. Fifth, this study is the first cohort study to confirm the relationship between combined UPPP and tonsillectomy and DNI, so it is hard to compare our result to previous research. Furthermore, the number of patients referred from ED and admitted to ICU was too small. Hence, the data were just for information. Last, much research on tonsillectomy and its impact on immunity in children has been done, but previous publications rarely mention the influence of tonsillectomy in adulthood. Since our study excluded patients aged below 18, whether tonsillectomy affects the adult immune system still needs more investigation to clarify.

5. Conclusions

Our study investigated the association between combined UPPP and tonsillectomy and DNI. Based on our study, we identified that combined UPPP and tonsillectomy does not increase the risk of DNI within 1–60 months. Moreover, it may imply the efficacy of combined UPPP and tonsillectomy in reducing the severity of DNI by decreasing the intubation rate and length of hospitalization.

Author Contributions: P.-C.H., L.-C.S. and Y.-A.T. contributed to designing the method, and wrote the first draft of the report, with input from the other authors. W.-D.C., J.-N.L., P.-S.L., C.-J.T., C.-D.L., H.-T.Y. and T.-C.S. collected data and conducted the data analyses. All authors have read and agreed to the published version of the manuscript.

Funding: This research received no funding.

Institutional Review Board Statement: The study protocol was approved by the Internal Review Board of China Medical University Hospital (CMUH104-REC2-115(CR-5)).

Informed Consent Statement: Not applicable.

Data Availability Statement: Data from the Taiwan National Health Insurance Research Database (NHIRD).

Acknowledgments: All authors thank the China Medical University (DMR-106-040 and DMR-110-205) and China Medical University Hospital (DMR-106-039 and DMR-106-198) in Taiwan for their support of this study.

Conflicts of Interest: The authors declare no conflict of interest.

References

1. Almuqamam, M.; Gonzalez, F.J.; Kondamudi, N.P. Deep Neck Infections. In *StatPearls*; StatPearls Publishing: Treasure Island, FL, USA, 2021.
2. Vieira, F.; Allen, S.M.; Stocks, R.M.; Thompson, J.W. Deep neck infection. *Otolaryngol. Clin. North Am.* **2008**, *41*, 459. [CrossRef] [PubMed]
3. Chiner, E.; Llombart, M.; Valls, J.; Pastor, E.; Sancho-Chust, J.N.; Andreu, A.L.; Sanchez-De-La-Torre, M.; Barbé, F. Association between Obstructive Sleep Apnea and Community-Acquired Pneumonia. *PLoS ONE* **2016**, *11*, e0152749. [CrossRef] [PubMed]
4. Eckert, D.J.; Malhotra, A. Pathophysiology of Adult Obstructive Sleep Apnea. *Proc. Am. Thorac. Soc.* **2008**, *5*, 144–153. [CrossRef] [PubMed]
5. Levy, P.; Pépin, J.L.; Arnaud, C.; Tamisier, R.; Borel, J.C.; Dematteis, M.; Godin-Ribuot, D.; Ribuot, C. Intermittent hypoxia and sleep-disordered breathing: Current concepts and perspectives. *Eur. Respir. J.* **2008**, *32*, 1082–1095. [CrossRef]
6. Lattimore, J.-D.L.; Celermajer, D.S.; Wilcox, I. Obstructive sleep apnea and cardiovascular disease. *J. Am. Coll. Cardiol.* **2003**, *41*, 1429–1437. [CrossRef]
7. Antonopoulou, S.; Loukides, S.; Papatheodorou, G.; Roussos, C.; Alchanatis, M. Airway inflammation in obstructive sleep apnea: Is leptin the missing link? *Respir. Med.* **2008**, *102*, 1399–1405. [CrossRef]
8. Inancli, H.M.; Enoz, M. Obstructive sleep apnea syndrome and upper airway inflammation. *Recent Patents Inflamm. Allergy Drug Discov.* **2010**, *4*, 54–57. [CrossRef]
9. Wang, X.; Ouyang, Y.; Wang, Z.; Zhao, G.; Liu, L.; Bi, Y. Obstructive sleep apnea and risk of cardiovascular disease and all-cause mortality: A meta-analysis of prospective cohort studies. *Int. J. Cardiol.* **2013**, *169*, 207–214. [CrossRef]
10. Eckert, D.J.; Malhotra, A.; Jordan, A.S. Mechanisms of apnea. *Prog. Cardiovasc. Dis.* **2009**, *51*, 313–323.
11. Sheen, D.; Abdulateef, S. Uvulopalatopharyngoplasty. *Oral Maxillofac. Surg. Clin. North Am.* **2021**, *33*, 295–303. [CrossRef]
12. Kinoshita, H.; Shibano, A.; Sakoda, T.; Ikeda, H.; Iranami, H.; Hatano, Y.; Enomoto, T. Uvulopalatopharyngoplasty decreases levels of C-reactive protein in patients with obstructive sleep apnea syndrome. *Am. Heart J.* **2006**, *152*, 692.e1–692.e5. [CrossRef] [PubMed]

13. Lee, H.M.; Kim, H.Y.; Suh, J.D.; Han, K.-D.; Kim, J.K.; Lim, Y.C.; Hong, S.-C.; Cho, J.H. Uvulopalatopharyngoplasty reduces the incidence of cardiovascular complications caused by obstructive sleep apnea: Results from the national insurance service survey 2007–2014. *Sleep Med.* **2018**, *45*, 11–16. [CrossRef] [PubMed]

14. Spicuzza, L.; Caruso, D.; Di Maria, G. Obstructive sleep apnoea syndrome and its management. *Ther. Adv. Chronic Dis.* **2015**, *6*, 273–285. [CrossRef] [PubMed]

15. Madani, M. Snoring and obstructive sleep apnea. *Arch. Iran. Med.* **2007**, *10*, 215–226.

16. Senska, G.; Ellermann, S.; Ernst, S.; Lax, H.; Dost, P. Recurrent tonsillitis in adults: Quality of life after tonsillectomy. *Dtsch. Arztebl. Int.* **2010**, *107*, 622–628.

17. Treviño-Gonzalez, J.L.; Maldonado-Chapa, F.; González-Larios, A.; Angel, J.A.M.-D.; Soto-Galindo, G.A.; García-Villanueva, J.M.Z. Deep Neck Infections: Demographic and Clinical Factors Associated with Poor Outcomes. *ORL J. Oto-Rhino-Laryngol. Its Relat.* **2022**, *84*, 130–138. [CrossRef]

18. Knapp, S. Diabetes and Infection: Is There a Link?—A Mini-Review. *Gerontology* **2013**, *59*, 99–104. [CrossRef]

19. Adoviča, A.; Veidere, L.; Ronis, M.; Sumeraga, G. Deep neck infections: Review of 263 cases. *Otolaryngol. Polska* **2017**, *71*, 37–42. [CrossRef]

20. Zamiri, B.; Hashemi, S.B.; Hashemi, S.H.; Rafiee, Z.; Ehsani, S. Prevalence of odontogenic deep head and neck spaces infection and its correlation with length of hospital stay. *J. Dent.* **2012**, *13*, 29–35.

21. Zhan, X.; Li, L.; Wu, C.; Chitguppi, C.; Huntley, C.; Fang, F.; Wei, Y. Effect of uvulopalatopharyngoplasty (UPPP) on atherosclerosis and cardiac functioning in obstructive sleep apnea patients. *Acta Oto-Laryngologica* **2019**, *139*, 793–797. [CrossRef]

22. Cho, J.H.; Suh, J.D.; Han, K.-D.; Jung, J.-H.; Lee, H.M. Uvulopalatopharyngoplasty May Reduce the Incidence of Dementia Caused by Obstructive Sleep Apnea: National Insurance Service Survey 2007–2014. *J. Clin. Sleep Med.* **2018**, *14*, 1749–1755. [CrossRef] [PubMed]

23. Marcano-Acuña, M.E.; Carrasco-Llatas, M.; Tortajada-Girbés, M.; Dalmau-Galofre, J.; Codoñer-Franch, P. Impact of adenotonsillectomy on the evolution of inflammatory markers. *Clin. Otolaryngol.* **2019**, *44*, 983–988. [CrossRef]

24. Wang, Y.-P.; Wang, M.-C.; Lin, H.-C.; Lee, K.-S.; Chou, P. Tonsillectomy and the Risk for Deep Neck Infection—A Nationwide Cohort Study. *PLoS ONE* **2015**, *10*, e0117535. [CrossRef] [PubMed]

25. Kaygusuz, I.; Alpay, H.C.; Gödekmerdan, A.; Karlidag, T.; Keles, E.; Yalcin, S.; Demir, N. Evaluation of long-term impacts of tonsillectomy on immune functions of children: A follow-up study. *Int. J. Pediatr. Otorhinolaryngol.* **2009**, *73*, 445–449. [CrossRef] [PubMed]

26. Altwairqi, R.G.; Aljuaid, S.M.; Alqahtani, A.S. Effect of tonsillectomy on humeral and cellular immunity: A systematic review of published studies from 2009 to 2019. *Eur. Arch. Otorhinolaryngol.* **2020**, *277*, 1–7. [CrossRef]

27. Bitar, M.A.; Dowli, A.; Mourad, M. The effect of tonsillectomy on the immune system: A systematic review and meta-analysis. *Int. J. Pediatr. Otorhinolaryngol.* **2015**, *79*, 1184–1191. [CrossRef]

28. Duval, M.; Daniel, S.J. Retropharyngeal and parapharyngeal abscesses or phlegmons in children. *Int. J. Pediatr. Otorhinolaryngol.* **2008**, *72*, 1765–1769. [CrossRef]

29. Byars, S.G.; Stearns, S.C.; Boomsma, J.J. Association of Long-Term Risk of Respiratory, Allergic, and Infectious Diseases with Removal of Adenoids and Tonsils in Childhood. *JAMA Otolaryngol. Neck Surg.* **2018**, *144*, 594–603. [CrossRef]

30. Nizam, N.; Basoglu, O.K.; Tasbakan, M.S.; Lappin, D.F.; Buduneli, N. Is there an association between obstructive sleep apnea syndrome and periodontal inflammation? *Clin. Oral Investig.* **2016**, *20*, 659–668. [CrossRef]

31. Price, R.; Kang, P. Association Between Periodontal Disease and Obstructive Sleep Apnea: What the Periodontist Should Know. *Compend. Contin. Educ. Dent.* **2020**, *41*, 149–153.

32. Xu, H.; Li, X.; Zheng, X.; Xia, Y.; Fu, Y.; Li, X.; Qian, Y.; Zou, J.; Zhao, A.; Guan, J.; et al. Pediatric Obstructive Sleep Apnea is Associated with Changes in the Oral Microbiome and Urinary Metabolomics Profile: A Pilot Study. *J. Clin. Sleep Med.* **2018**, *14*, 1559–1567. [CrossRef] [PubMed]

33. Ding, M.-C.; Hsu, C.-M.; Liu, S.; Lee, Y.-C.; Yang, Y.-H.; Liu, C.-Y.; Chang, G.-H.; Tsai, Y.-T.; Lee, L.-A.; Yang, P.-R.; et al. Deep Neck Infection Risk in Patients with Sleep Apnea: Real-World Evidence. *Int. J. Environ. Res. Public Health* **2021**, *18*, 3191. [CrossRef] [PubMed]

34. Sakarya, E.U.; Kulduk, E.; Gündoğan, O.; Soy, F.K.; Dündar, R.; Kılavuz, A.E.; Özbay, C.; Eren, E.; İmre, A. Clinical features of deep neck infection: Analysis of 77 patients. *J. Ear Nose Throat* **2015**, *25*, 102–108. [CrossRef]

35. Kauffmann, P.; Cordesmeyer, R.; Troltzsch, M.; Sommer, C.; Laskawi, R. Deep neck infections: A single-center analysis of 63 cases. *Med. Oral Patol. Oral. Cir. Bucal.* **2017**, *22*, e536–e541. [CrossRef]

36. O'Brien, K.J.; Bs, K.R.S.; Dugan, A.J.; Westgate, P.M.; Gupta, N. Risk Factors Affecting Length of Stay in Patients with Deep Neck Space Infection. *Laryngoscope* **2020**, *130*, 2133–2137. [CrossRef] [PubMed]

37. Hsu, Y.-S.; Hsu, W.-C.; Ko, J.-Y.; Yeh, T.-H.; Lee, C.-H.; Kang, K.-T. Incidence of Multilevel Surgical Procedures and Readmissions in Uvulopalatopharyngoplasty in Taiwan. *JAMA Otolaryngol. Neck Surg.* **2019**, *145*, 803–810. [CrossRef]

Review

Unraveling the Complexities of Oxidative Stress and Inflammation Biomarkers in Obstructive Sleep Apnea Syndrome: A Comprehensive Review

Salvatore Lavalle [1], Edoardo Masiello [2], Giannicola Iannella [3], Giuseppe Magliulo [3], Annalisa Pace [3], Jerome Rene Lechien [4], Christian Calvo-Henriquez [5], Salvatore Cocuzza [6], Federica Maria Parisi [6], Valentin Favier [7], Ahmed Yassin Bahgat [8], Giovanni Cammaroto [9], Luigi La Via [10], Caterina Gagliano [1], Alberto Caranti [11], Claudio Vicini [11] and Antonino Maniaci [1,*]

1. Faculty of Medicine and Surgery, University of Enna Kore, 94100 Enna, Italy; salvatore.lavalle@unikore.it (S.L.); caterina_gagliano@hotmail.com (C.G.)
2. Clinical and Experimental Radiology Unit, Experimental Imaging Center, IRCCS San Raffaele Scientific Institute, Via Olgettina 60, 20132 Milan, Italy; edo.masiello@gmail.com
3. Department of 'Organi di Senso', University "Sapienza", Viale dell'Università, 33, 00185 Rome, Italy; giannicola.iannella@uniroma1.it (G.I.); giuseppe.magliulo@uniroma1.it (G.M.); annalisa.pace@uniroma1.it (A.P.)
4. Department of Human Anatomy and Experimental Oncology, Faculty of Medicine, UMONS Research Institute for Health Sciences and Technology, University of Mons, 7022 Mons, Belgium; jerome.lechien@unimons.ac.be
5. Service of Otolaryngology, Hospital Complex of Santiago de Compostela, 15705 Santiago de Compostela, Spain; christian.calvo.henriquez@gmail.com
6. Department of Medical and Surgical Sciences and Advanced Technologies "GF Ingrassia", ENT Section, University of Catania, Via S. Sofia, 78, 95125 Catania, Italy; s.cocuzza@unict.it (S.C.); federicamariaparisi@gmail.com (F.M.P.)
7. Service d'ORL et de Chirurgie Cervico-Faciale, Centre Hospitalo-Universitaire de Montpellier, 80 Avenue Augustin Fliche, 34000 Montpellier, France
8. Department of Otorhinolaryngology, Alexandria University, Alexandria 21577, Egypt; ahmedyassinbahgat@gmail.com
9. Department of Head-Neck Surgery, Otolaryngology, Head-Neck and Oral Surgery Unit, Morgagni Pierantoni Hospital, Via Carlo Forlanini, 34, 47121 Forlì, Italy; giovanni.cammaroto@hotmail.com
10. Department of Anaesthesia and Intensive Care, University Hospital Policlinico-San Marco, 95125 Catania, Italy; luigilavia7@gmail.com
11. ENT and Audiology Department, University of Ferrara, 44121 Ferrara, Italy; dott.albertocaranti@gmail.com (A.C.); claudio@claudiovicini.com (C.V.)
* Correspondence: tnmaniaci209@gmail.com; Tel.: +39-3204-1545-76

Abstract: Background: Obstructive sleep apnea syndrome (OSAS), affecting approximately 1 billion adults globally, is characterized by recurrent airway obstruction during sleep, leading to oxygen desaturation, elevated carbon dioxide levels, and disrupted sleep architecture. OSAS significantly impacts quality of life and is associated with increased morbidity and mortality, particularly in the cardiovascular and cognitive domains. The cyclic pattern of intermittent hypoxia in OSAS triggers oxidative stress, contributing to cellular damage. This review explores the intricate relationship between OSAS and oxidative stress, shedding light on molecular mechanisms and potential therapeutic interventions. Methods: A comprehensive review spanning from 2000 to 2023 was conducted using the PubMed, Cochrane, and EMBASE databases. Inclusion criteria encompassed English articles focusing on adults or animals and reporting values for oxidative stress and inflammation biomarkers. Results: The review delineates the imbalance between pro-inflammatory and anti-inflammatory factors in OSAS, leading to heightened oxidative stress. Reactive oxygen species biomarkers, nitric oxide, inflammatory cytokines, endothelial dysfunction, and antioxidant defense mechanisms are explored in the context of OSAS. OSAS-related complications include cardiovascular disorders, neurological impairments, metabolic dysfunction, and a potential link to cancer. This review emphasizes the potential of antioxidant therapy as a complementary treatment strategy. Conclusions: Understanding the molecular intricacies of oxidative stress in OSAS is crucial for developing targeted therapeutic interventions. The comprehensive analysis of biomarkers provides insights into the complex interplay between OSAS and systemic complications, offering avenues for future research and therapeutic advancements in this multifaceted sleep disorder.

Keywords: obstructive sleep apnea syndrome (OSAS); intermittent hypoxia (IH); oxidative stress; inflammation biomarkers; reactive oxygen species (ROS); nitric oxide (NO); inflammatory cytokines; endothelial dysfunction; antioxidant defense; cellular damage

1. Introduction

Obstructive sleep apnea syndrome (OSAS) is a widespread and intricate respiratory disorder affecting nearly 1 billion adults aged 30–69 years globally, contingent upon geographical variations [1]. Characterized by recurrent episodes of upper airway obstruction during sleep, OSAS sets off a cascade of physiological events, leading to compromised oxygen saturation, elevated carbon dioxide levels, and recurrent arousals that disrupt sleep architecture [2,3]. This phenomenon results in a range of symptoms, including daytime somnolence, impaired cognitive function, and chronic fatigue, significantly impacting affected individuals' quality of life. Moreover, OSAS is a recognized contributor to morbidity and mortality, elevating the risk of cardiovascular pathologies [4,5], hypertension, cognitive dysfunction, and an accelerated aging process [6]. The cyclic pattern of intermittent hypoxia in OSAS triggers arterial chemoreceptors, heightening sympathetic nervous system activity [7]. This, in turn, influences vascular reactivity, contributing to the generation of free radicals—highly reactive molecules that interact with nucleic acids, proteins, and lipids, thereby altering cellular metabolism and causing cellular damage. This phenomenon, termed oxidative stress, represents an imbalance between the production of oxygen free radicals and antioxidant capacity, measurable through various biomarkers [8,9]. In addition to oxidative stress, OSAS induces pro-inflammatory factors, leading to the production of cytokines like tumor necrosis factor and interleukins 6 and 8 [9]. These cytokines are implicated in the pathogenesis of atherosclerosis and hypertension, positioning OSAS as an independent risk factor for these conditions [10]. While studies suggest an excess of reactive oxygen species in OSAS, consensus is lacking regarding specific markers to measure and the choice of antioxidants for mitigating the detrimental oxidative effects [11,12]. This review explores the intricate relationship between oxidative stress and OSAS, exploring the impact of intermittent hypoxia on the redox balance and the potential downstream effects on cellular and systemic health. By examining the current literature on oxidative stress in OSAS patients, we seek to shed light on the molecular mechanisms involved and the implications for therapeutic interventions targeting oxidative stress in this sleep disorder.

2. Materials and Methods

Study Protocol

A comprehensive review of the medical literature from January 2000 to December 2023 was conducted using databases such as PubMed, Cochrane, and EMBASE. We considered several study types, including clinical, preclinical, animal research, ongoing clinical trials, and literature reviews. We considered full-text English articles focusing on the adult population or animal subjects, providing reported values for at least one oxidative stress or inflammation marker.

The literature search was performed using a combination of key terms specific to the domains of obstructive sleep apnea and oxidative stress. Studies exploring inflammation biomarkers such as protein C reactive, tumor necrosis factor-alpha (TNF-α), interleukin 6 (IL-6), interleukin 8 (IL-8), NADPH oxidase, nitric oxide (NO), asymmetric dimethylarginine (ADMA), arginase, antioxidant system, glutathione, vitamin C, and vitamin E were retrieved. These carefully chosen keywords were pivotal in identifying studies pertinent to the intricate relationship between obstructive sleep apnea and markers indicative of oxidative stress and inflammation.

3. Results

This comprehensive review included 16 research articles in the final analysis. These articles explored several different biomarkers implicated in oxidative stress in patients with obstructive sleep apnea, aimed at proposing new biological markers useful in quantifying systemic inflammation related to OSA. As can be seen in the table below, the included studies were conducted on markers of reactive oxygen species, nitric oxide, inflammatory cytokines, antioxidant defense, and endothelial and organ dysfunction (Table 1). The inherent imbalance between pro-inflammatory and anti-inflammatory factors precipitates heightened oxidative stress, primarily attributed to an upsurge in oxygen free radicals coupled with an inadequate antioxidant capacity [13] (Figure 1). In the complex pathophysiology of obstructive sleep apnea syndrome (OSAS), the intricate interplay of molecular mechanisms begins with the activation of Hypoxia-Inducible Factor 1-alpha (HIF-1α) and nuclear factor kappa-light-chain-enhancer of activated B cells (NF-κB) in response to the chronic intermittent hypoxia that characterizes this condition [14–18]. Such activation is a pivotal adaptive response to hypoxemia, but it becomes maladaptive when repeatedly triggered, leading to a cascade of subsequent events [19–21]. The stabilization and activation of HIF-1α upregulate various genes, including those involved in oxidative stress and inflammatory responses, while NF-κB plays a crucial role in the transcriptional activation of pro-inflammatory cytokines [22]. The oscillating oxygen levels drive the generation of reactive oxygen species (ROS), such as superoxide dismutase (SOD), glutathione reduced (GSH), and catalase (CAT), overwhelming the endogenous antioxidant defenses and tipping the balance towards a state of oxidative stress [17]. This oxidative stress, in turn, facilitates the activation of NF-κB, further promoting the release of pro-inflammatory cytokines like tumor necrosis factor (TNF) and interleukins (IL), such as IL-6 and IL-8 [12,18]. These cytokines contribute to systemic inflammation and play a role in the development of endothelial dysfunction, a precursor to atherosclerosis and cardiovascular disease [8,23]. Elevated levels of these markers correlate with the severity of OSAS, typically measured by the apnea–hypopnea index (AHI). Furthermore, the depletion of antioxidant molecules like GSH and the accumulation of oxidized equivalents such as glutathione oxidized (GSSG) reflect the impaired redox state in OSAS patients. The intracellular ratio of GSSG to GSH rises, indicating oxidative stress, while the activity of enzymes like SOD and CAT is often found to be altered, reflecting the body's attempt to counteract the increased oxidative burden [22]. The results highlight a complex network of interrelated pathways involving HIF-1α and NF-κB activation, ROS production, antioxidant defense compromise, inflammatory cytokine release, and endothelial dysfunction, all contributing to the pathophysiological landscape of OSAS [11,17]. These mechanisms serve as both potential biomarkers for the severity of the disease and targets for therapeutic intervention to alleviate the systemic consequences of OSAS [15].

A comprehensive understanding of chronic systemic inflammation involves the quantification of various inflammatory biomarkers present in blood or urine, emanating from nucleic acids, proteins, and lipids. Furthermore, the recurrent cycles of chronic hypoxia/reoxygenation and sleep fragmentation, culminating in an augmented production of reactive oxygen species, circulating cytokines, and adhesion molecules, have been extensively correlated in the literature with cardiovascular, metabolic, and neurodegenerative comorbidities in individuals with OSAS. The interplay of these factors offers insights into the intricate connections between the physiological perturbations associated with OSAS and the development of associated health complications (Table 1).

Table 1. Properties of oxidative stress indicators in individuals with obstructive sleep apnea syndrome (OSAS). Abbreviation: SOD, superoxide dismutase; GSH, glutathione reduced; GSSG, glutathione oxidized; AHI, apnea–hypopnea index; TNF, tumor necrosis factor; NADPH, nicotinamide adenine dinucleotide phosphate; IL, interleukin; CAT, catalase.

Authors	Study Characteristics	Outcome
Reactive Oxygen Species		
Liu H.G., Zhou Y.N., Liu K. et al. (2010) [17]	30 OSAS patients vs. 23 healthy controls.	The mRNA levels of NADPH oxidase p22phox in sputum samples significantly increased in individuals with OSAS ($p < 0.05$).
R. Schulz, S. Mahmoudi, K. Hattarm et al. (2000) [21]	18 OSAS patients vs. two control groups of 10 healthy volunteers and 10 patients without OSAS.	The release of superoxide demonstrated a marked increase in each comparison ($p < 0.01$).
Nitric Oxide		
Duchna H.W., Guilleminault C., Stoohs R.A. et al. [23]	23 male OSAS patients and 12 male healthy controls.	Patients with OSAS exhibit impaired endothelium-dependent NO-mediated vasodilation ($p < 0.001$).
Kapusuz Gencer Z., Özkiriş M., Göçmen Y. et al. [24]	36 OSAS patients vs. 22 healthy controls.	There is a positive correlation between plasma NO levels and AHI.
Canino B., Hopps E., Calandrino V. et al. [25]	48 OSAS patients vs. 31 healthy controls.	Across the entire OSAS subject group, no significant difference in NO was identified when compared to the control group.
Wu, Z.H., Tang, Y., Niu, X. et al. [26]	Metanalysis of a total of 7 eligible studies, including 250 OSAS patients and 158 non-OSAS patients).	OSAS exhibited a significant association with serum or plasma NO levels (WMD = -11.66, 95% CI -17.21 to -6.11; $p < 0.01$), indicating that serum or plasma NO levels in OSAS patients are lower than those in controls.
Inflammatory Cytokines		
Lin C.C., Liaw S.F., Chiu C.H. et al. (2016) [27]	35 patients with moderately severe to severe OSAS vs. 20 healthy controls	TNF-α levels were elevated ($p < 0.01$).
Li X., Hu R., Ren X., He J. (2021) [28]	Metanalysis of a total of 25 eligible studies, including 2301 participants and 1123 controls to evaluate the association between serum IL-8 concentration and OSAS.	Correlation between serum IL-8 concentration and OSAS, revealed that both adults and children with OSAS exhibited higher serum IL-8 concentrations compared to controls (SMD = 0.997, 95% CI = 0.437–1.517, $p < 0.001$; SMD = 0.431, 95% CI = 0.104–0.759, $p = 0.01$).
Ifergane G., Ovanyan A., Toledano R. et al. (2016) [29]	The final analysis incorporated 43 individuals experiencing acute stroke and sleep apnea.	There was a positive correlation between AHI and IL-6 ($\rho = 0.37$, $p = 0.02$).
Wu M.F., Chen Y.H., Chen H.C. et al. (2020) [30]	The final analysis incorporated 100 participants, comprising 63 individuals with normal to moderate OSAS and 37 with severe OSAS.	There was a significant interaction effect on IL-6 levels for all OSAS severity and sex ($p = 0.030$). Additionally, IL-6 levels were higher in the obese group than in the non-obese group, irrespective of OSAS severity and sex ($p = 0.000$).
Yokoe T., Minoguchi K., Matsuo H. et al. (2003) [14]	A total of 30 individuals diagnosed with OSAS and 14 obese participants serving as control subjects.	IL-6 levels were significantly higher in patients with OSAS compared to the control group ($p < 0.05$).

Table 1. *Cont.*

Authors	Study Characteristics	Outcome
	Antioxidant Defense	
Tian Z., Sun H., Kang J. et al. (2022) [31]	Metanalysis of a total of 14 eligible studies, including 1240 patients and 457 controls.	The circulating SOD levels in patients with OSAS were significantly lower than those in the control group (SMD = −1.645, 95% CI = −2.279 to −1.011, $p < 0.001$).
Ntalapascha M., Makris D., Kyparos A. et al. (2012) [32]	18 patients with severe OSAS and 13 controls were included in the study.	The overnight ratio of GSH/GSSG and the levels of GSH were significantly different from controls ($p = 0.03$ and $p = 0.048$, respectively).
Sales L.V., Bruin V.M., D'Almeida V. et al. (2013) [33]	14 patients with obstructive sleep apnea vs. 13 controls.	Vitamin E exhibited lower levels ($p < 0.006$), SOD showed a decrease ($p < 0.001$), vitamin B11 demonstrated a decline ($p < 0.001$), and homocysteine concentrations increased ($p < 0.02$). Serum concentrations of vitamin C, CAT, GSH, and vitamin B12 remained unaltered.
Simiakakis M., Kapsimalis F., Chaligiannis E. et al. (2012) [18]	66 total subjects were referred (42 patients with OSAS vs. 24 controls).	The antioxidant capacity levels in OSAS patients were significantly lower ($p = 0.004$).
Mancuso M., Bonanni E., Lo Gerfo A. et al. (2012) [15]	41 untreated patients with a new diagnosis of OSAS vs. 32 healthy subjects.	The total GSH levels were significantly lower in OSAS patients than controls (95% CI for the mean 0.389–0.449 nmol/μL vs. 0.574–0.713 nmol/μL; $p < 0.0001$).

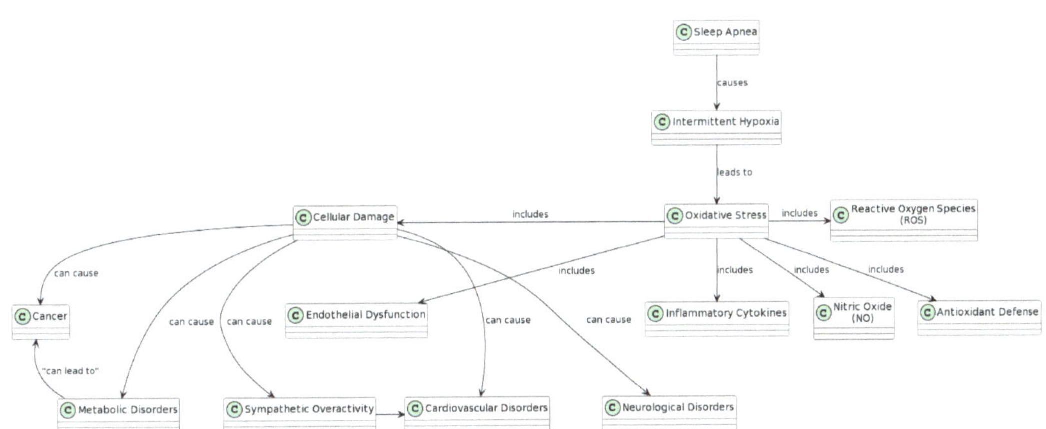

Figure 1. The figure illustrates the intricate network of oxidative stress and inflammation biomarkers in individuals with OSAS. The diagram categorizes the main biomarkers into physiopathogenetic clusters, providing a comprehensive overview of the molecular mechanisms involved in OSAS-related complications. Different clusters represent specific pathways, such as reactive oxygen species biomarkers, nitric oxide regulation, inflammatory cytokines, endothelial dysfunction, antioxidant defense, and cellular damage. The interconnections and associations between these clusters are visually depicted, offering a clear understanding of how OSAS induces oxidative stress, inflammation, and subsequent health complications.

3.1. Altered Sleep Architecture and Intermittent Hypoxia in Obstructive Sleep Apnea

Sleep architecture in OSA patients is significantly disrupted due to frequent awakenings or micro-awakenings and chronic intermittent hypoxia (CIH) typical of obstructive sleep apnea [4,34–36]. In patients with OSA, CIH and frequent awakenings lead to a significant reduction in the quantity and quality of N3 and REM sleep [37–39]. This disruption results in a sleep pattern characterized by excessive transitions between sleep stages, with more time spent in the lighter stages of sleep and less time in the deeper, restorative stages [40–42]. Molecularly, the sleep disruptions can impact the expression and regulation of various neurotransmitters and neural pathways that are critical for maintaining sleep stages, particularly the deeper stages of non-REM sleep and REM sleep [43,44]. The lack of restorative sleep further exacerbates the systemic effects of CIH, as it impairs the body's healing and metabolic processes that are typically more active during these deeper sleep stages. The interplay between CIH-induced molecular pathways and disrupted sleep architecture leads to a cycle of physiological stress and impaired tissue function [45–47]. OSA-induced sleep fragmentation sets off a domino effect of sympathetic nervous activation, circadian rhythm disruption, inflammatory pathway engagement, endocrine dysregulation, and oxidative stress, all of which intertwine to contribute to the multi-system impact of this sleep disorder [48–53]. The interrelated nature of these mechanisms highlights the importance of addressing sleep quality and architecture in the management and treatment of OSA. Altered sleep architecture in obstructive sleep apnea leads to a complex cascade of pathophysiological events, particularly due to sleep fragmentation, starting with the activation of the sympathetic nervous system [50–52]. Each arousal catapults the body into a 'fight or flight' state, increasing heart rate and blood pressure, which, over time, can result in cardiovascular complications and heightened stress responses [53–55]. Simultaneously, the disrupted sleep pattern wreaks havoc on the body's circadian rhythms. These rhythms are essential for regulating not only sleep and wakefulness but also various hormonal outputs such as melatonin and cortisol [5,24,30]. The disarray caused by OSA can lead to mood disturbances and metabolic issues as these hormones become dysregulated. Moreover, as sleep is fragmented, the stages of sleep that usually help to down-regulate pro-inflammatory pathways are interrupted, resulting in elevated levels of inflammatory cytokines like IL-6 and TNF-α [53–57]. This state of chronic inflammation is a contributing factor to systemic health issues and can accelerate atherosclerotic processes [58–60]. The repercussions of fragmented sleep and heightened sympathetic activity also extend to endocrine functions. The normal secretion patterns of hormones, including those from the hypothalamic-pituitary-adrenal axis and growth hormone, are altered, contributing to an array of metabolic dysregulations, such as insulin resistance and abnormal appetite control [24]. Furthermore, the oxidative stress burden increases as the body's antioxidant defenses are compromised due to inadequate restorative sleep, leading to cellular damage and contributing to the risk of developing cardiovascular disease, neurodegeneration, and other oxidative stress-related pathologies [33,43,44]. The association of Hypoxia-Inducible Factor 1-alpha (HIF-1α) with the pathophysiological processes of obstructive sleep apnea (OSA) is a critical element [10]. HIF-1α is instrumental in the body's adaptive response to the hypoxic conditions characteristic of OSA, which result from repeated airway blockages leading to intermittent hypoxia. Upon a decrease in oxygen levels, HIF-1α stabilizes and accumulates, triggering the transcription of various genes aimed at helping the body adjust to the lack of oxygen [20]. HIF-1α is deeply involved in the molecular pathways that lead to oxidative stress, sympathetic overactivity, and systemic inflammation seen in patients with OSA. It influences the expression of genes associated with angiogenesis, erythropoiesis, and glucose metabolism, and it escalates the production of reactive oxygen species (ROS) by enhancing the expression of enzymes responsible for mitochondrial respiration [22]. This escalation in ROS may exceed the capacity of antioxidant defenses, causing oxidative stress, which subsequently damages DNA, proteins, and lipids within cells [21,26,30]. Additionally, HIF-1α plays a role in the inflammatory response that is characteristic of OSA. It can initiate the transcription of pro-inflammatory cytokines and adhesion molecules, which contribute to systemic inflammation and endothelial dysfunction [40,51]. These mechanisms are pivotal in the onset of cardiovascular diseases, which are frequently seen as comorbid conditions in individuals with OSA. Recognizing the significance of HIF-1α in the pathogenesis of OSA is vital, for it may highlight new therapeutic targets. Modulating HIF-1α signaling pathways could potentially reduce the negative impact of intermittent hypoxia on oxidative stress and inflammation.

3.2. Reactive Oxygen Species Biomarkers

Reactive oxygen species (ROS) biomarkers in OSAS play a crucial role in unraveling the intricate relationship between this sleep disorder and the oxidative stress it induces. The respiratory or oxidative burst orchestrates the generation and release of ROS, including the superoxide anion, hydrogen peroxide, hydroxyl radical, and singlet oxygen [14].

Various stimuli, such as the characteristic hypoxia observed in OSAS, can initiate this reaction [15]. Nicotinamide adenine dinucleotide phosphate oxidase (NADPH oxidase), an enzyme, plays a central role in this process, converting free oxygen (O_2) into superoxide and subsequently triggering the production of other reactive molecules like hydroxide anions, peroxide, hypochlorite, and nitrogen monoxide [16,17]. Importantly, this enzymatic step is responsible for oxidizing biological compounds such as lipids, proteins, and DNA, leading to altered plasma concentrations of associated oxidative markers [18]. Elevated levels of the superoxide anion, a primary ROS, are observed in OSAS. This molecule is a key contributor to oxidative stress, participating in various pathways that can lead to cellular damage [19]. Superoxide anions are also converted to hydrogen peroxide by superoxide dismutase enzymes; while hydrogen peroxide is less reactive and can function as a signaling molecule, in excessive amounts, it contributes to oxidative damage, necessitating the activation of cellular defenses like catalase and glutathione peroxidase for its breakdown. Hydrogen peroxide (H_2O_2), another ROS, is produced because of the dismutation of the superoxide anion. Increased H_2O_2 levels in OSAS contribute to oxidative stress, affecting cellular components and signaling pathways [20]. The hydroxyl radical ($\bullet OH$), one of the most potent ROS, is generated through the Fenton reaction. Its presence in OSAS signifies a heightened state of oxidative stress, potentially impacting cellular structures and functions. Singlet oxygen ($1O_2$), a highly reactive form of oxygen, can initiate oxidative damage to biomolecules such as lipids, proteins, and DNA [21]. These reactive species are proficient at initiating lipid peroxidation, which compromises cell membranes and creates toxic byproducts that further damage macromolecules. Proteins, vital for myriad cellular functions, are susceptible to structural and functional alterations due to the oxidative modification of amino acids, with consequences that include disrupted enzyme activities and signaling pathways. DNA is not spared; it undergoes oxidative attacks that can result in mutations or even genomic instability, potentially leading to cell death or carcinogenesis. Mitochondria, implicated in both ROS generation and targeting, suffer from oxidative damage that impairs their function, culminating in a dysfunctional energy supply and the release of signals that promote cell death. Specific studies directly addressing singlet oxygen in OSAS are limited but, along with other ROS, may contribute to endothelial dysfunction, inflammation, and tissue damage observed in OSAS patients [22]. ROS-induced endothelial dysfunction can precipitate a series of events that restrict blood flow and further deprive tissues of oxygen, setting the stage for more ROS production. Compounding the problem, ROS can activate matrix metalloproteinases that degrade the extracellular matrix, undermining tissue architecture and stability. The balance between cell survival and death is also tipped, as apoptosis and necrosis pathways are triggered by excessive ROS levels, contributing to organ dysfunction. Even autophagy, a cellular cleanup process, can be thrown into disarray by ROS, leading to the accumulation of cellular debris and dysfunction. Moreover, ROS influence cell signaling pathways, sometimes fostering pathological conditions by promoting aberrant cell proliferation and survival.

3.3. Nitric Oxide

Nitric oxide (NO) assumes a multifaceted role in OSAS. Synthesized by nitric oxide synthase (NOS), thanks to the essential amino acid L-arginine. NO regulates vascular tone through vasodilation, but its balance is influenced by OSAS-related factors like intermittent hypoxia. OSAS is associated with endothelial dysfunction, and the reduced bioavailability of NO, often linked to oxidative stress and inflammation, may contribute to cardiovascular complications, the mediating vasodilating effect, and platelet aggregation [4,33]. The interaction between NO and ROS in the context of intermittent hypoxia can lead to the formation of peroxynitrite by interaction with superoxide or the action of dimethylarginine (ADMA), potentially contributing to oxidative stress. At elevated levels, it disrupts the synthesis of NO by diminishing the activity of the enzyme dimethylarginine dimethylaminohydrolase, resulting in increased levels of ADMA [12,45–48]. In recent times, several investigations have delved into the levels of NOx in individuals with OSAHS. Nevertheless, the findings are inconclusive. Kapusuz et al. [24] observed notably elevated plasma NO levels in OSAS patients in contrast to the control group, whereas Canino et al. [25] did not observe any distinction in NO levels between OSAS subjects and their healthy counterparts. A recent meta-analysis conducted by Wu et al. found that OSAS was significantly related to serum or plasma NO levels and that serum or plasma NO levels in OSAS patients are lower than in controls [30]. Studies suggest a connection between OSAS, diminished NO bioavailability, and increased cardiovascular risks, with impaired NO-mediated vasodilation contributing to hypertension and atherosclerosis [4,32,33]. NO may also play a role in sleep regulation, and changes in its levels could contribute to the sleep disturbances observed in OSAS [38]. Despite its involvement in various physiological processes, the precise mechanisms and therapeutic implications of NO in OSAS remain areas of active research.

3.4. Antioxidant Defense

The antioxidant defense system includes various components such as enzymes (e.g., superoxide dismutase, catalase, and peroxidase) and non-enzymatic molecules (e.g., glutathione, vitamin C, and vitamin E) [13,22]. These antioxidants work synergistically to neutralize ROS and prevent oxidative damage to cellular components like lipids, proteins, and DNA. Individuals with OSAS exhibit an imbalance between the production of oxidative agents and the compensatory action performed by the antioxidant system, known as total antioxidant capacity (TAC) [7,18,39,40]. Superoxide, a vital cellular oxidizing agent, undergoes dismutation catalyzed by the superoxide dismutase (SOD) enzyme family, leading to the dissociation of the superoxide anion into molecular oxygen and hydrogen peroxide in healthy individuals [41–45]. However, OSAS patients have been reported to have lower plasma levels of SOD [30,31]. Catalase (CAT) plays a crucial role in mitigating oxidative stress by facilitating the breakdown of hydrogen peroxide into water and molecular oxygen [39]. The primary function of glutathione peroxidase (GPx) is to catalyze the reduction of hydrogen peroxide (H_2O_2) and organic hydroperoxides, utilizing reduced glutathione (GSH) as a substrate [45]. This enzymatic reaction helps prevent the accumulation of harmful ROS within cells and tissues [40]. The glutathione system, which includes GPx, acts as a first line of defense against oxidative damage by neutralizing peroxides and maintaining the cellular redox balance [49]. A study conducted by Asker et al. demonstrates that OSAS patients had lower levels of CAT and GPX [46]. Interestingly, a correlation was detected between CAT and GPX levels and polysomnographic indices. Both correlated directly with the AHI, but glutathione peroxidase levels were inversely correlated with the mean duration of apnea. Literature suggests that molecules like glutathione, vitamin C, and vitamin E contribute to ameliorating oxidative stress in OSAS patients, especially in conjunction with continuous positive airway pressure (CPAP) therapy [49,55]. Additionally, oxidative stress is linked to sleep disturbances in OSAS patients, and the intake of antioxidants has been shown to enhance sleep quality [7]. Sales et al. discovered reduced antioxidants in OSAS patients, indicating a correlation between antioxidants and neuropsychological alterations in obstructive sleep apnea [33]. Specifically, they observed decreased levels of vitamin E ($p < 0.006$), superoxide dismutase ($p < 0.001$), and vitamin B11 ($p < 0.001$), along with increased homocysteine levels ($p < 0.02$).

3.5. Inflammatory Cytokines

In individuals affected by OSAS, encompassing both children and adults, the presence of chronic intermittent hypoxia induces a systemic inflammatory response. Additionally, key inflammatory cytokines such as interleukin-8, tumor necrosis factor-alpha, and interleukin-6 exhibit upregulation, potentially linked to the activation of the nuclear factor pathway [46]. OSAS is characterized as a low-grade chronic inflammatory respiratory condition, as the repetitive episodes of chronic intermittent hypoxia during sleep instigate an anti-inflammatory cascade. These cytokines, often elevated in the serum of OSAS patients, can exert widespread effects on multiple organ systems. The upregulation of these cytokines in OSAS is thought to be closely associated with the activation of the nuclear factor kappa-light-chain-enhancer of activated B cells (NF-κB) pathway. NF-κB is a transcription factor that plays a critical role in the inflammatory response. Under normal conditions, NF-κB is sequestered in the cytoplasm by the inhibitor IκB. However, during episodes of hypoxia, IκB is phosphorylated and degraded, freeing NF-κB to translocate into the nucleus, where it can initiate the transcription of various inflammatory genes, including those for IL-8, TNF-α, and IL-6. These inflammatory mediators contribute to the pathophysiology of OSAS by promoting leukocyte recruitment, inducing the expression of adhesion molecules on endothelial cells, and elevating the production of reactive oxygen species (ROS). Moreover, pro-inflammatory cytokines may modulate metabolic processes, influence the hepatic production of acute-phase reactants, and contribute to the development of insulin resistance.

3.5.1. Tumor Necrosis Factor-α

Tumor necrosis factor-alpha (TNF-α) serves as a pivotal proinflammatory cytokine in the intricate immunological landscape of OSAS. OSAS, characterized by recurrent episodes of hypoxia and reoxygenation, manifests an augmented inflammatory milieu marked by heightened systemic levels of TNF-α. Evidence suggests that TNF-α levels positively correlate with the severity of OSAS [47]. The nuanced interplay of TNF-α with other inflammatory mediators in OSAS contributes to the complexity of the inflammatory cascade, with cumulative effects implicated in the development of cardiovascular comorbidities commonly observed in individuals with OSAS. TNF-α plays a significant role in promoting atherosclerosis, inducing the expression of cellular adhesion molecules, and facilitating the adhesion of leukocytes to the vascular endothelium [48,49]. Elevated circulating levels of TNF-α have been correlated with early atherosclerotic signs in healthy middle-aged individuals [46,50–53]. Furthermore, these levels serve as predictive markers for coronary heart disease and congestive cardiac failure. In the context of OSAS, TNF-α concentration is elevated compared to healthy subjects, underscoring its potential contribution to cardiovascular risks in OSAS patients. Remarkably, continuous positive airway pressure (CPAP) treatment demonstrates the capacity to normalize TNF values in OSAS individuals, suggesting a potential avenue for mitigating the inflammatory impact associated with this sleep disorder [29–31].

3.5.2. Interleukin-8

Interleukin-8 (IL-8), acknowledged as one of the most potent inflammatory cell chemokines, plays a crucial role in initiating systemic inflammation in OSAS and associated cardiovascular conditions. IL-8 functions by inducing myeloperoxidase release from neutrophils and recruiting inflammatory cells, contributing to a sustained inflammatory response. Akyol et al. reported that IL-8 binding to specific receptors on neutrophil surfaces leads to cell deformation, degranulation, and increased production of reactive oxygen species [53]. This process may activate arachidonic acid through lysosomal secretion, resulting in heightened vascular permeability, plasma protein exudation, and subsequent tissue damage, atherosclerosis, and other diseases [54]. Recent meta-analysis findings emphasize that individuals, both children and adults, with OSAS exhibit significantly elevated IL-8 concentrations, with IL-8 levels positively correlating with the severity of OSAS indicated by the apnea–hypopnea index (AHI) and being linked to obesity and ethnicity [55].

3.5.3. Interleukin-6

Interleukin-6 (IL-6) is a multifunctional cytokine with several biological activities, such as the proliferation of T lymphocytes, the differentiation of B lymphocytes, and the stimulation of immunoglobulin secretion [51,58]. Moreover, IL-6 plays a role in regulating the natural sleep patterns associated with the circadian secretion pattern. In individuals with OSAS, particularly during episodes of intermittent hypoxia and reoxygenation, there is an upregulation of IL-6 as part of the systemic inflammatory response [59–62]. A meta-analysis performed by Nadeem et al. indicates elevated levels of interleukin-6 (IL-6) in patients with OSAS compared to control individuals [46]. Furthermore, a recent metanalysis performed by Imani et al. confirms the significant correlation between IL-6 and the AHI, indicating a potential link between IL-6 and the severity of OSAS, and has also highlighted a positive correlation between IL-6 production and body mass index [59].

3.6. Endothelial Dysfunction

A potential early sign of vascular disease is endothelial dysfunction [60]. Research indicates that individuals with OSAS who have not experienced vascular issues previously exhibit endothelial dysfunction [61]. Unfortunately, the specific mechanisms triggering the development of endothelial dysfunction in OSA remain unclear [62]. Exposure to harmful cellular risks, such as oxidative stress, may result in endothelial dysfunction [63–65], leading to a reduction in its ability to dilate blood vessels, an elevation in proinflammatory and prothrombotic reactions, and abnormal regulation of vascular growth [66]. OSAS induces intermittent hypoxia, triggering oxidative stress and inflammation. This, in turn, prompts the release of proinflammatory cytokines, such as IL-6 and TNF-α, and elevates C-reactive protein levels (CRP) [46,55]. These inflammatory mediators lead to an endothelium with proinflammatory tendencies and subsequent endothelial dysfunction. This dysfunctional state is characterized by an elevation in the expression of cell adhesion molecules (CAMs), including E-selectin, vascular cell adhesion molecule-1 (VCAM-1), and intercellular adhesion molecule-1 (ICAM-1) [67,68]. These molecules facilitate the adhesion and migration of leukocytes into the vessel wall, a crucial step in the initiation of atherosclerosis [69,70]. A dysfunctional endothelium also exhibits diminished nitric oxide levels, a reduction that may result from an elevated CRP level, leading to the downregulation of endothelial nitric oxide synthase (eNOS) expression and bioactivity [71]. Furthermore, the reduction in NO levels may be attributed to the presence of superoxide anions arising from an imbalance between ROS synthesis and antioxidant systems, leading to oxidative stress [72]. These pathophysiological mechanisms associated with OSAS potentially contribute to the development of cardiovascular events, such as systemic hypertension and other cardiovascular diseases, and are a key factor in developing atherosclerotic plaques [73].

3.7. Cellular Damage and Organ Dysfunction

In individuals with OSAS, repetitive airway collapse and obstruction due to various pathological factors result in recurrent apnea, periodic arousal during sleep, intermittent hypoxia (IH), and sleep fragmentation. These core processes trigger various cellular [71] and molecular mechanisms, including increased sympathetic nerve activity [69], metabolic dysregulation [58], systemic inflammation [73], oxidative stress, and endothelial dysfunction [60]. These mechanisms, identified as pathogenic in clinical and experimental models, contribute to OSAS-related complications across different systems [61–64].

3.7.1. Cardiovascular Disorders

Notably, OSAS is strongly associated with cardiovascular complications, including systemic hypertension, arrhythmias, coronary artery disease, and stroke [50,57,66,73]. The link between OSAS and hypertension is particularly significant, with up to 80% of patients with resistant hypertension potentially suffering from OSAS [73]. The increase in sympathetic nerve activity, driven by ROS, is a prominent feature of OSAS and is implicated in OSAS-related cardiovascular issues [74,75]. Oxidative stress, inflammation, and molecular mechanisms play crucial roles in developing cardiocerebrovascular diseases in OSAS patients [76]. In addition, intermittent hypoxia (IH) and recurrent arousals, likely through mechanisms involving oxidative stress and activation of Hypoxia-Inducible Factor

1 (HIF-1), result in sympathetic overactivity in patients with obstructive sleep apnea (OSA). The effects of this overactivity include elevated catecholamine levels, systemic hypertension, changes in ventricular repolarization, and cardiac remodeling. These physiological changes contribute to the cardiovascular burden often observed in individuals with OSA [60–64]. The complex interplay of these factors underscores the importance of early intervention and treatment of breathing disorders during sleep to prevent cardiovascular morbidity [77]. In addition, the recurrent changes in intrathoracic pressure associated with obstructive sleep apnea (OSA) provoke an augmented venous return to the heart, which in turn can cause an overload of the right ventricle. Moreover, the intrathoracic pressure dips below that of the external structures surrounding the heart, which increases the afterload on the left ventricle. This heightened afterload can impair the heart's systolic and diastolic functions. Over time, these pressures may lead to a chronic dilation of the left atrium, which could have significant implications for cardiac health and function. Animal studies have provided significant insights into the cardiovascular consequences of obstructive sleep apnea (OSA) and the molecular mechanisms underlying these effects. In various animal models, OSA is simulated through induced intermittent hypoxia, mirroring the oxygen desaturation-reoxygenation cycles seen in human OSA. These studies have shown that such hypoxic episodes can lead to sympathetic nervous system activation, oxidative stress, and systemic inflammation—all factors contributing to cardiovascular pathology. Key molecular pathways include activation of the sympathetic nervous system and downstream signaling processes such as those mediated by HIF-1α, which have been implicated in the development of hypertension and atherosclerosis. In rodent models, intermittent hypoxia has been shown to lead to endothelial dysfunction, vascular remodeling, and a propensity for arrhythmogenesis, providing a mechanistic basis for the association between OSA and increased cardiovascular risk. In humans, the relationship between OSA and cardiovascular events has been extensively studied through randomized clinical trials and observational cohorts. The evidence suggests that OSA is independently associated with an increased risk of hypertension, coronary artery disease, heart failure, and arrhythmias, most notably atrial fibrillation. continuous positive airway pressure (CPAP) therapy has been the cornerstone of OSA management, with several trials demonstrating its efficacy in reducing apneic events and improving the quality of sleep. However, its role in the secondary prevention of cardiovascular events remains controversial. While some studies have shown that CPAP treatment can lower blood pressure and reduce the risk of recurrent cardiovascular events, others have not found a significant benefit in terms of cardiovascular outcomes. This has led to an ongoing debate in the field, with some experts suggesting that the heterogeneity in patient populations, varying adherence to CPAP treatment, and differences in baseline cardiovascular risk may contribute to these conflicting results. Future studies with rigorous design, perhaps focusing on personalized medicine approaches to identify those most likely to benefit from CPAP, are needed to clarify its role in cardiovascular risk reduction among patients with OSA.

3.7.2. Neurological Disorders

Prolonged exposure to IH in patients with OSAS has profound effects on various central nervous system (CNS) functions, resulting in severe neurocognitive and behavioral deficits. OSAS is associated with a decline in cognitive functions, including memory, executive function, and comprehension, as well as mood disturbances, insomnia, and excessive daytime sleepiness [77,78]. Animal studies indicate that IH induces neuronal injury, inflammation, and astrocyte activation in the rat brain, leading to impaired cognitive performance in tasks such as the Morris water maze test [78,79]. Clinical studies in OSAS patients reveal cognitive impairments in attention, delayed memory function, and executive function, which are correlated with the severity of OSAS. Structural and functional alterations in brain anatomy, including decreased gray matter in various regions, provide indirect evidence of brain damage in OSAS patients [76,80]. The brain, being sensitive to hypoxia, experiences oxidative stress, inflammation, and neuronal damage due to IH. The involvement of ROS, oxidative stress, inflammatory damage, and microglial activation contributes to neuronal apoptosis and/or necrosis, leading to OSAS-related cognitive impairments. The NF-κB, TNF-α, CRP, IL-1β, IL-6, and cyclooxygenase-2 (COX-2) pathways are implicated in neuroinflammation and cognitive dysfunction in OSAS [78,79]. Microglia, as major inflammatory cells in the CNS, mediate oxidative stress and inflammation, and their activation is associated with neurotoxicity [80]. The NF-κB-mediated JNK and p38 MAPK pathways play crucial roles in hippocampal injury and cognitive dysfunction [17]. Additionally, brain-derived neurotrophic factor (BDNF) and excitotoxic neurotransmitters such as glutamate contribute to OSAS-related CNS damage [81]. The accumulating evidence highlights the intricate relationship between inflammation and cognitive impairment in OSAS, suggesting potential links with neurological disorders that warrant further investigation.

3.7.3. OSAS, Obesity, and Metabolic Disorders

Emerging evidence from animal models of OSAS suggests that IH is independently linked to metabolic dysfunction. OSAS demonstrates an independent association with insulin resistance, implying its potential role in the development of type 2 diabetes and metabolic syndrome, characterized by obesity, insulin resistance, hypertension, and dyslipidemia [82]. Clinical studies have revealed significantly higher levels of fasting blood glucose and insulin resistance in OSAS patients, with the severity of OSAS correlating with increased insulin resistance [83]. This association extends to non-obese patients, and AHI has been identified as an independent risk factor for insulin resistance and type 2 diabetes [84]. IH-induced oxidative stress and inflammation in OSAS contribute to insulin resistance, with inflammatory factors inhibiting insulin receptors and the phosphorylation of insulin receptor substrates. IH also impacts glucose metabolism by reducing glucose uptake in muscles, affecting pancreatic β-cell function, and increasing sympathetic tone, thereby disrupting glycemic and insulin homeostasis. OSAS is further implicated in lipid abnormalities, elevating total cholesterol, triglycerides, LDL, and VLDL levels [85].

Obstructive sleep apnea (OSA) is closely intertwined with obesity, a condition that itself is a well-established pro-inflammatory state. The high prevalence of overweight and obese individuals among OSA patients complicates the understanding of the systemic inflammation observed in OSA. Adipose tissue in obese individuals is not merely a storage depot for excess calories but an active endocrine organ that secretes a variety of cytokines and inflammatory mediators, such as TNF-α, IL-6, and C-reactive protein (CRP). These mediators contribute to chronic, low-grade systemic inflammation. In OSA, intermittent hypoxia and sleep fragmentation further exacerbate this inflammatory milieu. However, distinguishing the inflammation due to OSA from that due to obesity can be challenging, as both conditions independently contribute to systemic inflammation and share common pathophysiological pathways. As such, the inflammation observed in OSA patients may be compounded by the presence of obesity, making it a confounding factor in the assessment and management of inflammation in OSA. This overlap implies that the therapeutic strategies targeting OSA should also consider the management of obesity to effectively mitigate the compounded inflammatory state. Treatment with CPAP may positively influence lipid profiles. IH-associated changes in leptin and adiponectin levels contribute to insulin sensitivity and metabolic homeostasis [86–88]. Although the exact relationship between OSAS and metabolic diseases is still debated, recognizing their strong association is crucial for early detection and intervention. Further research is needed to elucidate specific mechanisms and address controversies in this complex relationship.

3.7.4. OSAS, Oxidative Stress, and Cancer

In recent years, accumulating circumstantial, epidemiological, clinical, and experimental evidence has strongly suggested a notable impact of OSAS on tumorigenesis and tumor development. A comprehensive multicenter cohort study involving cancer-free OSAS patients revealed a significant association between nocturnal hypoxemia and overall cancer incidence [89]. Moreover, individuals under 45 years old with severe OSAS demonstrated a markedly increased risk of various cancer types compared to the general population [90]. Epidemiological investigations further confirmed a link between OSAS and elevated cancer-related mortality, revealing a dose–response relationship between OSAS severity and cancer-specific mortality. This association spans over a 22-year follow-up period, where severe OSAS was associated with nearly a fivefold risk of death from cancer [91–94]. OSAS is implicated in raising the incidence of specific tumor types, including lung cancer, breast cancer, prostate cancer, nasopharyngeal tumors, and melanoma. Notably, in certain tumors, exposure to IH, mimicking the oxygenation pattern induced by OSAS during sleep, has been shown to promote the growth, invasion, and metastasis of lung cancer, colon cancer, and melanoma [92].

3.8. OSA Treatment Effectiveness on Inflammation and Oxidative Stress

CPAP remains the gold standard treatment for OSA. Numerous studies have demonstrated the effectiveness of CPAP in reducing systemic inflammation, a key player in the pathogenesis of atherosclerosis and cardiovascular disease [95–98]. Inflammatory biomarkers, such as C-reactive protein (CRP), TNF-α, and interleukins (IL-6 and IL-8), have been shown to decrease significantly with compliant use of CPAP therapy [99]. Oxidative stress, which contributes to endothelial dysfunction and subsequent cardiovascular disease, is also mitigated by CPAP [100]. Markers of oxidative stress, such as malondialdehyde (MDA) and nitric oxide (NO) levels, exhibit notable improvements with CPAP use [101]. The impact of CPAP on reducing inflammation and oxidative stress has profound implications for comorbid conditions. For instance, cardiovascular risk factors such as hypertension and arrhythmias are markedly improved with effective CPAP therapy, likely due to the reduction in sympathetic nervous system activity and improved vascular endothelial function [102]. Similarly, CPAP use has been linked to improvements in insulin sensitivity and lipid profiles, reducing the risk for metabolic syndrome and type 2 diabetes [103,104]. The reduction of inflammation and oxidative stress through the treatment of OSA has a favorable impact on several comorbid conditions [105]. For instance, the cardiovascular benefits of reducing these pathological processes are substantial, leading to a decrease in the incidence of myocardial infarction, stroke, and heart failure [106]. Additionally, improvements in metabolic outcomes, such as better glucose control and lipid metabolism, can significantly reduce the risk of diabetes and contribute to weight loss [107]. Cognitive benefits are also noteworthy, as untreated OSA is associated with an increased risk of cognitive decline and dementia. By reducing inflammation and oxidative stress, which are implicated in neurodegeneration, treatments for OSA may also preserve cognitive function and reduce the risk of neurocognitive disorders [108]. Other respiratory indices have been assessed to identify predictors of OSA treatment. The study by Fernandes et al. analyzed the relationship between mean oxygen saturation (SpO$_2$) and inflammatory markers in OSA patients treated with CPAP [109]. They found that subjects with a lower mean SpO$_2$ (<95%) had a higher inflammatory profile, including a higher number of leukocytes, a higher number of neutrophils, a higher number of basophils, and an elevated concentration of C-reactive protein. These results suggest that SpO$_2$ levels may play a role in predicting the inflammatory status and treatment outcome of OSAS subjects. Conversely, the study conducted by Ming et al. reported that TNF-α levels were negatively correlated with the mean and lowest oxygen saturation levels (MSaO$_2$ and LSaO$_2$). Additionally, they observed a positive correlation between IL-8 levels and AHI, as well as morning systolic and diastolic blood pressure, while IL-8 levels were negatively correlated with MSaO$_2$ and LSaO$_2$. These results suggest that TNF-α and IL-8 may be variably involved in the inflammatory and cardiovascular consequences of obstructive sleep apnea [110]. In the study conducted by Tauman et al., they found that children with moderate-severe sleep-disordered breathing (SDB) had increased plasma levels of IL-6 compared to controls, and this increase was statistically significant ($p = 0.03$). In particular, the levels of IL-6 were positively correlated with the apnea-hypopnea index (AHI) (r = 0.28, $p = 0.003$) and negatively correlated with the lowest oxygen saturation levels (SpO_2 nadir) (r = −0.24, *p* = 0.02). Additionally, the study revealed that children with SDB exhibited severity-dependent increases in plasma C-reactive protein (CRP) and IL-6 levels, regardless of their obesity status. Although less frequently, outcomes on oxidative stress, inflammation, and different specific biomarkers have been evaluated in other types of treatment. MADs are oral appliances designed to advance the mandible and, consequently, the base of the tongue, enlarging the airway space to reduce apneic events [111]. While CPAP is more effective in reducing the apnea–hypopnea index (AHI), MADs offer a viable alternative for patients with mild to moderate OSA or those who are non-compliant with CPAP. Studies have shown that MADs can lead to improvements in inflammatory markers similar to those seen with CPAP, albeit to a lesser extent in some cases [112]. The reduction in oxidative stress with MAD use, while still a topic of ongoing research, has promising preliminary results, suggesting that they can offer cardiovascular protective effects [113,114]. Surgical options for OSA aim to address anatomical abnormalities contributing to airway obstruction. These procedures range from uvulopalatopharyngoplasty (UPPP) to more complex surgeries such as maxillomandibular advancement (MMA) [34,35]. The impact of surgery on inflammatory and oxidative stress markers is less clear than with CPAP or MADs, primarily due to the variability in surgical techniques and individual patient anatomy [115]. However, successful surgical outcomes that lead to a significant reduction in AHI do correlate with a decrease in systemic inflammation and oxidative stress [116]. Despite the potential benefits, surgical treatments are often considered a last resort due to their invasive nature and associated risks. When surgery successfully reduces or eliminates apneic events, it can have a significant impact on reducing the overall inflammatory burden and the risk of cardiovascular disease [117]. Moreover, surgery may provide a permanent solution for selected patients, which can be particularly appealing compared to the need for ongoing treatment with CPAP or MADs.

3.9. Future Perspectives for Sleep Apnea Biomarkers

The burgeoning field of biomarker research in OSA is uncovering novel pathways and targets that could revolutionize the diagnosis and treatment of this complex disorder. Fan et al. have shown that NAD+ biosynthesis reduction may lead to mitochondrial dysfunction and vascular endothelial injury, which are critical in the pathogenesis of OSA [63]. Under chronic intermittent hypoxia (CIH), the study found a decrease in NAD+ biosynthesis due to inhibited NAMPT enzyme activity, which led to mitochondrial dysfunction in endothelial cells, characterized by reduced ATP and mitochondrial membrane potential, impaired respiratory chain activity, increased oxidative stress, and compromised vascular function. Supplementing with nicotinamide mononucleotide (NMN) reversed the mitochondrial and endothelial dysfunction caused by CIH. However, endothelial damage induced by oxidized low-density lipoprotein (ox-LDL) did not show involvement of the NAD+ pathway and was not mitigated by NMN supplementation. Similarly, Chen et al. have identified that the long, non-coding RNA FKSG29 plays a pivotal role in regulating oxidative stress and endothelial dysfunction in OSA, suggesting a new molecular target for intervention [64]. The authors reported that FKSG29 and certain pro-oxidant genes were upregulated in OSA patients, while anti-oxidant genes were downregulated compared to primary snorers. In vitro, knocking down FKSG29 in cells exposed to intermittent hypoxia with re-oxygenation (IHR) reduced reactive oxygen species production, apoptosis, and abnormal gene expression associated with oxidative stress, and these protective effects were negated by concurrently knocking down miR-23a-3p. The research suggested that targeting the FKSG29/miR-23a-3p/IL6R pathway could be a novel therapeutic strategy for OSA-induced endothelial dysfunction. Promising data are also present on the role of miRNA as a novel biomarker for OSA patients. The work of Fadaei et al. supports the potential utility of circulating miRNAs, specifically miR125a, miR126, and miR146a-5p, as biomarkers for endothelial dysfunction in OSA patients, which could serve as non-invasive diagnostic tools [65]. The transition from physiological adaptation to pathological maladaptation in response to chronic intermittent hypoxia, a hallmark of OSA, is discussed by Arnaud et al., providing insights into the systemic effects of the disorder [66]. Nguyen et al. highlighted the role of peripheral inflammation due to sleep fragmentation, suggesting that inflammation biomarkers could be a key to understanding and treating OSA-related comorbidities [93]. In particular, the authors found that acute sleep fragmentation in male C57BL/6J mice induced a swift activation of the hypothalamic-pituitary-adrenal axis, increasing serum corticosterone levels within 1 h and persisting up to 24 h. Instead, a peripheral inflammatory response was evidenced by elevated pro-inflammatory gene expression in the heart from 1 h of ASF and a delayed increase in serum IL-6 concentration after 6 h. Collectively, a future can be envisioned where a comprehensive biomarker panel derived from genetic, molecular, and inflammatory markers could be developed, facilitating a more nuanced approach to OSA management, enabling earlier diagnosis, better risk stratification, and more precise targeting of therapies to ameliorate the multifactorial consequences of the disorder.

4. Study Limitations

Notwithstanding the fact that this study offers a thorough synthesis of the most recent data about biomarkers in OSAS, some limitations must be noted. One major drawback is that many of these biomarkers are not measured using uniform methods and assays across studies, which could explain part of the variation in reported findings. Few studies used longitudinal follow-up, and the majority had sample sizes that were quite small. The cohorts' limited applicability to women and other age groups resulted from their predominance of middle-aged male participants. Confounding variables were frequently not sufficiently taken into consideration, including underlying comorbidities, drugs, and lifestyle choices. Synthesizing biomarker data is further complicated by the variety of OSAS populations, which vary in severity, symptoms, and co-occurring medical illnesses. To validate findings, more high-quality research with reliable techniques, larger sample numbers, confounder adjustment, and longitudinal evaluations are warranted. Assay and biomarker panel standardization would enable cross-study comparability. It is necessary to include a variety of patient demographics that reflect the heterogeneity of OSAS in the real world. Although this analysis offers a solid foundation, its shortcomings point to the need for additional thorough investigation to improve our knowledge of the mechanisms of inflammation and oxidative stress that underlie the pathophysiology and clinical consequences of OSAS. Filling up these gaps would increase the possibility of biomarkers being translated into better diagnosis, prognosis, and treatment for OSAS.

5. Conclusions

Obstructive sleep apnea syndrome (OSAS) significantly impacts health, driving a range of complications through oxidative stress and inflammation. This disorder affects multiple physiological processes, contributing to cardiovascular, neurological, and metabolic disorders, and may increase cancer risk. The pathophysiology of OSA involves a complex interaction of factors, including sympathetic activation, endothelial dysfunction, hypoxia-induced metabolic imbalance, and increased inflammatory and proatherogenic activity. The intermittent hypoxia characteristic of OSAS leads to oxidative stress, systemic inflammation, and subsequent multi-organ dysfunction. Cardiovascular issues, cognitive decline, and metabolic syndrome are closely linked to OSAS, necessitating multidisciplinary research and clinical approaches. The potential role of OSAS in cancer progression also highlights the need for further sleep-related oncology research. Despite progress, gaps remain, particularly in identifying biomarkers and effective antioxidant therapies. Understanding OSAS's molecular mechanisms is critical for developing targeted treatments and integrating sleep medicine into comprehensive patient care.

Author Contributions: Conceptualization, S.L. and A.M.; methodology, E.M.; software, G.I.; validation, G.M., A.P. and A.Y.B.; formal analysis, S.C. and J.R.L.; investigation, C.C.-H.; resources, C.G.; data curation, S.C.; writing—original draft preparation, F.M.P. and C.V.; writing—review and editing, V.F.; visualization, L.L.V. and G.C.; supervision, A.C. and S.C.; project administration, A.M.; funding acquisition, S.L. All authors have read and agreed to the published version of the manuscript.

Funding: This research received no external funding.

Institutional Review Board Statement: Not applicable.

Informed Consent Statement: Informed consent was obtained from all subjects involved in the study.

Data Availability Statement: No new data were created or analyzed in this study. Data sharing is not applicable to this article.

Conflicts of Interest: The authors declare no conflicts of interest.

References

1. Benjafield, A.V.; Ayas, N.T.; Eastwood, P.R.; Heinzer, R.; Ip, M.S.; Morrell, M.J.; Nunez, C.M.; Patel, S.R.; Penzel, T.; Pépin, J.L.; et al. Estimation of the global prevalence and burden of obstructive sleep apnoea: A literature-based analysis. *Lancet Respir. Med.* **2019**, *7*, 687–698. [CrossRef]
2. Hu, Y.; Mai, L.; Luo, J.; Shi, W.; Xiang, H.; Song, S.; Hong, L.; Long, W.; Mo, B.; Luo, M. Peripheral blood oxidative stress markers for obstructive sleep apnea—A meta-analysis. *Sleep Breath.* **2022**, *26*, 2045–2057. [CrossRef]
3. Parish, J.M.; Somers, V.K. Obstructive sleep apnea and cardiovascular disease. *Mayo Clin. Proc.* **2004**, *79*, 1036–1046. [CrossRef] [PubMed]
4. Bradley, T.D.; Floras, J.S. Obstructive sleep apnoea and its cardiovascular consequences. *Lancet* **2009**, *373*, 82–93. [CrossRef] [PubMed]
5. Eisele, H.-J.; Markart, P.; Schulz, R. Obstructive Sleep Apnea, Oxidative Stress, and Cardiovascular Disease: Evidence from Human Studies. *Oxid. Med. Cell. Longev.* **2015**, *2015*, 608438. [CrossRef]
6. Li, Y.; Wang, Y. Obstructive Sleep Apnea-hypopnea Syndrome as a Novel Potential Risk for Aging. *Aging Dis.* **2021**, *12*, 586. [CrossRef]
7. Lira, A.B.; de Sousa Rodrigues, C.F. Evaluation of oxidative stress markers in obstructive sleep apnea syndrome and additional antioxidant therapy: A review article. *Sleep Breath.* **2016**, *20*, 1155–1160. [CrossRef] [PubMed]
8. Cofta, S.; Wysocka, E.; Piorunek, T.; Rzymkowska, M.; Batura-Gabryel, H.; Torlinski, L. Oxidative stress markers in the blood of persons with different stages of obstructive sleep apnea syndrome. *J. Physiol. Pharmacol.* **2008**, *59* (Suppl. S6), 183–190.
9. Peppard, P.E.; Young, T.; Palta, M.; Skatrud, J. Prospective Study of the Association between Sleep-Disordered Breathing and Hypertension. *N. Engl. J. Med.* **2000**, *342*, 1378–1384. [CrossRef]
10. Li, J.; Punjabi, N.M.; Sun, C.K.; Schwartz, A.R.; Smith, P.L.; Marino, R.L.; Rodriguez, A.; Hubbard, W.C.; O'Donnell, C.P.; Polotsky, V.Y. Intermittent Hypoxia Induces Hyperlipidemia in Lean Mice. *Circ. Res.* **2005**, *97*, 698–706. [CrossRef]
11. Lavie, L. Obstructive sleep apnoea syndrome—An oxidative stress disorder. *Sleep Med. Rev.* **2003**, *7*, 35–51. [CrossRef]
12. Sahlman, J.; Miettinen, K.; Peuhkurinen, K.; Seppä, J.; Peltonen, M.; Herder, C.; Punnonen, K.; Vanninen, E.; Gylling, H.; Partinen, M.; et al. The activation of the inflammatory cytokines in overweight patients with mild obstructive sleep apnoea. *J. Sleep Res.* **2010**, *19*, 341–348. [CrossRef]
13. Passali, D.; Corallo, G.; Yaremchuk, S.; Longini, M.; Proietti, F.; Passali, G.C.; Bellussi, L. Oxidative stress in patients with obstructive sleep apnoea syndrome. *Acta Otorhinolaryngol. Ital.* **2015**, *35*, 420–425. [CrossRef]
14. Yokoe, T.; Minoguchi, K.; Matsuo, H.; Oda, N.; Minoguchi, H.; Yoshino, G.; Hirano, T.; Adachi, M. Elevated Levels of C-Reactive Protein and Interleukin-6 in Patients With Obstructive Sleep Apnea Syndrome Are Decreased by Nasal Continuous Positive Airway Pressure. *Circulation* **2003**, *107*, 1129–1134. [CrossRef]
15. Mancuso, M.; Bonanni, E.; LoGerfo, A.; Orsucci, D.; Maestri, M.; Chico, L.; DiCoscio, E.; Fabbrini, M.; Siciliano, G.; Murri, L. Oxidative stress biomarkers in patients with untreated obstructive sleep apnea syndrome. *Sleep Med.* **2012**, *13*, 632–636. [CrossRef]
16. Katsoulis, K.; Kontakiotis, T.; Spanogiannis, D.; Vlachogiannis, E.; Kougioulis, M.; Gerou, S.; Daskalopoulou, E. Total antioxidant status in patients with obstructive sleep apnea without comorbidities: The role of the severity of the disease. *Sleep Breath.* **2011**, *15*, 861–866. [CrossRef] [PubMed]
17. Liu, F.; Liu, T.-W.; Kang, J. The role of NF-κB-mediated JNK pathway in cognitive impairment in a rat model of sleep apnea. *J. Thorac. Dis.* **2018**, *10*, 6921–6931. [CrossRef] [PubMed]
18. Simiakakis, M.; Kapsimalis, F.; Chaligiannis, E.; Loukides, S.; Sitaras, N.; Alchanatis, M. Lack of Effect of Sleep Apnea on Oxidative Stress in Obstructive Sleep Apnea Syndrome (OSAS) Patients. *PLoS ONE* **2012**, *7*, e39172. [CrossRef] [PubMed]
19. Han, Y.; Fan, Z.D.; Yuan, N.; Xie, G.Q.; Gao, J.; De, W.; Gao, X.Y.; Zhu, G.Q. Superoxide anions in the paraventricular nucleus mediate the enhanced cardiac sympathetic afferent reflex and sympathetic activity in renovascular hypertensive rats. *J. Appl. Physiol.* **2011**, *110*, 646–652. [CrossRef] [PubMed]

20. Lavie, L. Oxidative stress in obstructive sleep apnea and intermittent hypoxia—Revisited—The bad ugly and good: Implications to the heart and brain. *Sleep Med. Rev.* **2015**, *20*, 27–45. [CrossRef] [PubMed]

21. Schulz, R.; Mahmoudi, S.; Hattar, K.; Sibelius, U.; Olschewski, H.; Mayer, K.; Seeger, W.; Grimminger, F. Enhanced release of superoxide from polymorphonuclear neutrophils in obstructive sleep apnea. Impact of continuous positive airway pressure therapy. *Am. J. Respir. Crit. Care Med.* **2000**, *162*, 566–570. [CrossRef] [PubMed]

22. Maniaci, A.; Iannella, G.; Cocuzza, S.; Vicini, C.; Magliulo, G.; Ferlito, S.; Cammaroto, G.; Meccariello, G.; De Vito, A.; Nicolai, A.; et al. Oxidative Stress and Inflammation Biomarker Expression in Obstructive Sleep Apnea Patients. *J. Clin. Med.* **2021**, *10*, 277. [CrossRef] [PubMed]

23. Duchna, H.-W.; Guilleminault, C.; Stoohs, R.A.; Orth, M.; de Zeeuw, J.; Schultze-Werninghaus, G.; Rasche, K. Das obstruktive Schlafapnoe-Syndrom: Ein kardiovaskulärer Risikofaktor? *Z. Kardiol.* **2001**, *90*, 568–575. [CrossRef] [PubMed]

24. Kapusuz Gencer, Z.; Özkiriş, M.; Göçmen, Y.; Intepe, Y.S.; Akin, I.; Delibaş, N.; Saydam, L. The correlation of serum levels of leptin, leptin receptor and NOx (NO_2^- and NO_3^-) in patients with obstructive sleep apnea syndrome. *Eur. Arch. Otorhinolaryngol.* **2014**, *271*, 2943–2948. [CrossRef]

25. Canino, B.; Hopps, E.; Calandrino, V.; Montana, M.; Lo Presti, R.; Caimi, G. Nitric oxide metabolites and erythrocyte deformability in a group of subjects with obstructive sleep apnea syndrome. *Clin. Hemorheol. Microcirc.* **2015**, *59*, 45–52. [CrossRef] [PubMed]

26. Wu, Z.-H.; Tang, Y.; Niu, X.; Sun, H.-Y. The role of nitric oxide (NO) levels in patients with obstructive sleep apnea-hypopnea syndrome: A meta-analysis. *Sleep Breath.* **2021**, *25*, 9–16. [CrossRef] [PubMed]

27. Lin, C.-C.; Liaw, S.F.; Chiu, C.H.; Chen, W.J.; Lin, M.W.; Chang, F.T. Effects of nasal CPAP on exhaled SIRT1 and tumor necrosis factor-α in patients with obstructive sleep apnea. *Respir. Physiol. Neurobiol.* **2016**, *228*, 39–46. [CrossRef] [PubMed]

28. Li, X.; Hu, R.; Ren, X.; He, J. Interleukin-8 concentrations in obstructive sleep apnea syndrome: A systematic review and meta-analysis. *Bioengineered* **2021**, *12*, 10650–10665. [CrossRef]

29. Ifergane, G.; Ovanyan, A.; Toledano, R.; Goldbart, A.; Abu-Salame, I.; Tal, A.; Stavsky, M.; Novack, V. Obstructive Sleep Apnea in Acute Stroke. *Stroke* **2016**, *47*, 1207–1212. [CrossRef]

30. Wu, M.-F.; Chen, Y.-H.; Chen, H.-C.; Huang, W.-C. Interactions among Obstructive Sleep Apnea Syndrome Severity, Sex, and Obesity on Circulatory Inflammatory Biomarkers in Patients with Suspected Obstructive Sleep Apnea Syndrome: A Retrospective, Cross-Sectional Study. *Int. J. Environ. Res. Public Health* **2020**, *17*, 4701. [CrossRef]

31. Tian, Z.; Sun, H.; Kang, J.; Mu, Z.; Liang, J.; Li, M. Association between the circulating superoxide dismutase and obstructive sleep apnea: A meta-analysis. *Eur. Arch. Otorhinolaryngol.* **2022**, *279*, 1663–1673. [CrossRef]

32. Ntalapascha, M.; Makris, D.; Kyparos, A.; Tsilioni, I.; Kostikas, K.; Gourgoulianis, K.; Kouretas, D.; Zakynthinos, E. Oxidative stress in patients with obstructive sleep apnea syndrome. *Sleep Breath.* **2013**, *17*, 549–555. [CrossRef]

33. Sales, L.V.; Bruin, V.M.S.; D'Almeida, V.; Pompéia, S. Cognition and biomarkers of oxidative stress in obstructive sleep apnea. *Clinics* **2013**, *68*, 449–455. [CrossRef] [PubMed]

34. Iannella, G.; Magliulo, G.; Di Luca, M.; De Vito, A.; Meccariello, G.; Cammaroto, G.; Pelucchi, S.; Bonsembiante, A.; Maniaci, A.; Vicini, C. Lateral pharyngoplasty techniques for obstructive sleep apnea syndrome: A comparative experimental stress test of two different techniques. *Eur. Arch. Otorhinolaryngol.* **2020**, *277*, 1793–1800. [CrossRef] [PubMed]

35. Iannella, G.; Vallicelli, B.; Magliulo, G.; Cammaroto, G.; Meccariello, G.; De Vito, A.; Greco, A.; Pelucchi, S.; Sgarzani, R.; Corso, R.M. Long-Term Subjective Outcomes of Barbed Reposition Pharyngoplasty for Obstructive Sleep Apnea Syndrome Treatment. *Int. J. Environ. Res. Public Health* **2020**, *17*, 1542. [CrossRef] [PubMed]

36. Vgontzas, A.N.; Zoumakis, E.; Lin, H.M.; Bixler, E.O.; Trakada, G.; Chrousos, G.P. Marked Decrease in Sleepiness in Patients with Sleep Apnea by Etanercept, a Tumor Necrosis Factor-α Antagonist. *J. Clin. Endocrinol. Metab.* **2004**, *89*, 4409–4413. [CrossRef]

37. McNicholas, W.T. Obstructive Sleep Apnea and Inflammation. *Prog. Cardiovasc. Dis.* **2009**, *51*, 392–399. [CrossRef] [PubMed]

38. Haight, J.S.J.; Djupesland, P.G. Nitric Oxide (NO) and Obstructive Sleep Apnea (OSA). *Sleep Breath.* **2003**, *7*, 53–61. [CrossRef] [PubMed]

39. Borges, Y.G.; Cipriano, L.H.C.; Aires, R.; Zovico, P.V.C.; Campos, F.V.; de Araújo, M.T.M.; Gouvea, S.A. Oxidative stress and inflammatory profiles in obstructive sleep apnea: Are short-term CPAP or aerobic exercise therapies effective? *Sleep Breath.* **2020**, *24*, 541–549. [CrossRef] [PubMed]

40. Wang, F.; Liu, Y.; Xu, H.; Qian, Y.; Zou, J.; Yi, H.; Guan, J.; Yin, S. Association between Upper-airway Surgery and Ameliorative Risk Markers of Endothelial Function in Obstructive Sleep Apnea. *Sci. Rep.* **2019**, *9*, 20157. [CrossRef]

41. Cespuglio, R.; Amrouni, D.; Meiller, A.; Buguet, A.; Gautier-Sauvigné, S. Nitric oxide in the regulation of the sleep-wake states. *Sleep Med. Rev.* **2012**, *16*, 265–279. [CrossRef]

42. Zhou, L.; Chen, P.; Peng, Y.; Ouyang, R. Role of Oxidative Stress in the Neurocognitive Dysfunction of Obstructive Sleep Apnea Syndrome. *Oxid. Med. Cell. Longev.* **2016**, *2016*, 9626831. [CrossRef] [PubMed]

43. Iliams, R.; Lemaire, P.; Lewis, P.; McDonald, F.B.; Lucking, E.; Hogan, S.; Sheehan, D.; Healy, V.; O'Halloran, K.D. Chronic intermittent hypoxia increases rat sternohyoid muscle NADPH oxidase expression with attendant modest oxidative stress. *Front. Physiol.* **2015**, *6*, 15.

44. Wali, S.O.; Bahammam, A.S.; Massaeli, H.; Pierce, G.N.; Iliskovic, N.; Singal, P.K.; Kryger, M.H. Susceptibility of LDL to oxidative stress in obstructive sleep apnea. *Sleep* **1998**, *21*, 290–296. [PubMed]

45. Diaz-Vivanco, P.; de Simone, A.; Kiddle, G.; Foyer, C.H. Glutathione—Linking cell proliferation to oxidative stress. *Free Radic. Biol. Med.* **2015**, *89*, 1154–1164. [CrossRef] [PubMed]

46. Asker, S.; Asker, M.; Sarikaya, E.; Sunnetcioglu, A.; Aslan, M.; Demir, H. Oxidative stress parameters and their correlation with clinical, metabolic and polysomnographic parameters in severe obstructive sleep apnea syndrome. *Int. J. Clin. Exp. Med.* **2015**, *8*, 11449–11455. [PubMed]

47. İn, E.; Özdemir, C.; Kaman, D.; Sökücü, S.N. Concentraciones de proteínas de estrés térmico, L-arginina y dimetilarginina asimétrica en pacientes con síndrome de apnea obstructiva del sueño. *Arch. Bronconeumol.* **2015**, *51*, 544–550. [CrossRef] [PubMed]

48. Zinellu, A.; Fois, A.G.; Mangoni, A.A.; Paliogiannis, P.; Sotgiu, E.; Zinellu, E.; Marras, V.; Pirina, P.; Carru, C. Systemic concentrations of asymmetric dimethylarginine (ADMA) in chronic obstructive pulmonary disease (COPD): State of the art. *Amino Acids* **2018**, *50*, 1169–1176. [CrossRef]

49. Wysocka, E.; Cofta, S.; Cymerys, M.; Gozdzik, J.; Torlinski, L.; Batura-Gabryel, H. The impact of the sleep apnea syndrome on oxidant-antioxidant balance in the blood of overweight and obese patients. *J. Physiol. Pharmacol.* **2008**, *59* (Suppl. S6), 761–769.

50. Nadeem, R.; Molnar, J.; Madbouly, E.M.; Nida, M.; Aggarwal, S.; Sajid, H.; Naseem, J.; Loomba, R. Serum Inflammatory Markers in Obstructive Sleep Apnea A Meta-Analysis. *J. Clin. Sleep Med.* **2013**, *9*, 1003–1012.

51. Jurado-Gámez, B.; Fernandez-Marin, M.C.; Gómez-Chaparro, J.L.; Muñoz-Cabrera, L.; Lopez-Barea, J.; Perez-Jimenez, F.; Lopez-Miranda, J. Relationship of oxidative stress and endothelial dysfunction in sleep apnoea. *Eur. Respir. J.* **2011**, *37*, 873–879. [CrossRef]

52. Testelmans, D.; Tamisier, R.; Barone-Rochette, G.; Baguet, J.P.; Roux-Lombard, P.; Pépin, J.L.; Lévy, P. Profile of circulating cytokines: Impact of OSA, obesity and acute cardiovascular events. *Cytokine* **2013**, *62*, 210–216. [CrossRef]

53. Chen, C.-Y.; Chen, C.L.; Yu, C.C.; Chen, T.T.; Tseng, S.T.; Ho, C.H. Association of inflammation and oxidative stress with obstructive sleep apnea in ischemic stroke patients. *Sleep Med.* **2015**, *16*, 113–118. [CrossRef]

54. Arnardottir, E.S.; Maislin, G.; Schwab, R.J.; Staley, B.; Benediktsdottir, B.; Olafsson, I.; Juliusson, S.; Romer, M.; Gislason, T.; Pack, A.I. The Interaction of Obstructive Sleep Apnea and Obesity on the Inflammatory Markers C-Reactive Protein and Interleukin-6: The Icelandic Sleep Apnea Cohort. *Sleep* **2012**, *35*, 921–932. [CrossRef]

55. Campos-Rodriguez, F.; Asensio-Cruz, M.I.; Cordero-Guevara, J.; Jurado-Gamez, B.; Carmona-Bernal, C.; Gonzalez-Martinez, M.; Troncoso, M.F.; Sanchez-Lopez, V.; Arellano-Orden, E.; Garcia-Sanchez, M.I.; et al. Effect of continuous positive airway pressure on inflammatory, antioxidant, and depression biomarkers in women with obstructive sleep apnea: A randomized controlled trial. *Sleep* **2019**, *42*, sz145. [CrossRef]

56. Li, Q.; Zheng, X. Tumor necrosis factor alpha is a promising circulating biomarker for the development of obstructive sleep apnea syndrome: A meta-analysis. *Oncotarget* **2017**, *8*, 27616–27626. [CrossRef] [PubMed]

57. Akyol, S.; Çörtük, M.; Baykan, A.O.; Kiraz, K.; Börekçi, A.; Şeker, T.; Gür, M.; Çayli, M. Mean platelet volume is associated with disease severity in patients with obstructive sleep apnea syndrome. *Clinics* **2015**, *70*, 481–485. [CrossRef] [PubMed]

58. Hirano, T.; Akira, S.; Taga, T.; Kishimoto, T. Biological and clinical aspects of interleukin 6. Immunol. *Today* **1990**, *11*, 443–449.

59. Imani, M.M.; Sadeghi, M.; Khazaie, H.; Emami, M.; Sadeghi Bahmani, D.; Brand, S. Evaluation of Serum and Plasma Interleukin-6 Levels in Obstructive Sleep Apnea Syndrome: A Meta-Analysis and Meta-Regression. *Front. Immunol.* **2020**, *11*, 1343. [CrossRef] [PubMed]

60. Brevetti, G.; Silvestro, A.; Schiano, V.; Chiariello, M. Endothelial Dysfunction and Cardiovascular Risk Prediction in Peripheral Arterial Disease. *Circulation* **2003**, *108*, 2093–2098. [CrossRef] [PubMed]

61. Khayat, R.N.; Varadharaj, S.; Porter, K.; Sow, A.; Jarjoura, D.; Gavrilin, M.A.; Zweier, J.L. Angiotensin Receptor Expression and Vascular Endothelial Dysfunction in Obstructive Sleep Apnea. *Am. J. Hypertens.* **2018**, *31*, 355–361. [CrossRef] [PubMed]

62. Jelic, S.; Padeletti, M.; Kawut, S.M.; Higgins, C.; Canfield, S.M.; Onat, D.; Colombo, P.C.; Basner, R.C.; Factor, P.; LeJemtel, T.H. Inflammation, Oxidative Stress, and Repair Capacity of the Vascular Endothelium in Obstructive Sleep Apnea. *Circulation* **2008**, *117*, 2270–2278. [CrossRef] [PubMed]

63. Fan, Z.T.; Dong, L.P.; Niu, Y.H.; Chi, W.W.; Wu, G.L.; Song, D.M. Specific role of NAD+ biosynthesis reduction mediated mitochondrial dysfunction in vascular endothelial injury induced by chronic intermittent hypoxia. *Eur. Rev. Med. Pharmacol. Sci.* **2023**, *27*, 10749–10762. [PubMed]

64. Chen, Y.C.; Hsu, P.Y.; Su, M.C.; Chen, Y.L.; Chang, Y.T.; Chin, C.H.; Lin, I.C.; Chen, Y.M.; Wang, T.Y.; Lin, Y.Y.; et al. Long non-coding RNA FKSG29 regulates oxidative stress and endothelial dysfunction in obstructive sleep apnea. *Mol. Cell. Biochem.* **2023**. *online ahead of print*. [CrossRef] [PubMed]

65. Fadaei, R.; Fallah, S.; Moradi, M.T.; Rostampour, M.; Khazaie, H. Circulating levels of miR125a, miR126, and miR146a-5p in patients with obstructive sleep apnea and their relation with markers of endothelial dysfunction. *PLoS ONE* **2023**, *18*, e0287594. [CrossRef]

66. Arnaud, C.; Billoir, E.; de Melo Junior, A.F.; Pereira, S.A.; O'Halloran, K.D.; Monteiro, E.C. Chronic intermittent hypoxia-induced cardiovascular and renal dysfunction: From adaptation to maladaptation. *J. Physiol.* **2023**, *601*, 5553–5577. [CrossRef]

67. Pak, V.M.; Keenan, B.T.; Jackson, N.; Grandner, M.A.; Maislin, G.; Teff, K.; Schwab, R.J.; Arnardottir, E.S.; Júlíusson, S.; Benediktsdottir, B.; et al. Adhesion molecule increases in sleep apnea: Beneficial effect of positive airway pressure and moderation by obesity. *Int. J. Obes.* **2015**, *39*, 472–479. [CrossRef]

68. Nikitidou, O.; Daskalopoulou, E.; Papagianni, A.; Vlachogiannis, E.; Dombros, N.; Liakopoulos, V. The impact of OSA and CPAP treatment on cell adhesion molecules' night-morning variation. *Sleep Breath.* **2021**, *25*, 1301–1307. [CrossRef]

69. Medina-Leyte, D.J.; Zepeda-García, O.; Domínguez-Pérez, M.; González-Garrido, A.; Villarreal-Molina, T.; Jacobo-Albavera, L. Endothelial Dysfunction, Inflammation and Coronary Artery Disease: Potential Biomarkers and Promising Therapeutical Approaches. *Int. J. Mol. Sci.* **2021**, *22*, 3850. [CrossRef]

70. Zeite, A.R.; Borges-Canha, M.; Cardoso, R.; Neves, J.S.; Castro-Ferreira, R.; Leite-Moreira, A. Novel Biomarkers for Evaluation of Endothelial Dysfunction. *Angiology* **2020**, *71*, 397–410.

71. Pober, J.S.; Sessa, W.C. Evolving functions of endothelial cells in inflammation. *Nat. Rev. Immunol.* **2007**, *7*, 803–815. [CrossRef]

72. Sies, H. Oxidative stress: A concept in redox biology and medicine. *Redox Biol.* **2015**, *4*, 180–183. [CrossRef]

73. Durán-Cantolla, J.; Aizpuru, F.; Martínez-Null, C.; Barbé-Illa, F. Obstructive sleep apnea/hypopnea and systemic hypertension. *Sleep Med. Rev.* **2009**, *13*, 323–331. [CrossRef]

74. Khurana, S.; Sharda, S.; Saha, B.; Kumar, S.; Guleria, R.; Bose, S. Canvassing the aetiology, prognosis and molecular signatures of obstructive sleep apnoea. *Biomarkers* **2019**, *24*, 1–16. [CrossRef]

75. Lavie, L.; Lavie, P. Molecular mechanisms of cardiovascular disease in OSAHS: The oxidative stress link. *Eur. Respir. J.* **2009**, *33*, 1467–1484. [CrossRef]

76. Rosenzweig, I.; Glasser, M.; Polsek, D.; Leschziner, G.D.; Williams, S.C.; Morrell, M.J. Sleep apnoea and the brain: A complex relationship. *Lancet Respir. Med.* **2015**, *3*, 404–414. [CrossRef] [PubMed]

77. Lv, R.; Liu, X.; Zhang, Y.; Dong, N.; Wang, X.; He, Y.; Yue, H.; Yin, Q. Pathophysiological mechanisms and therapeutic approaches in obstructive sleep apnea syndrome. *Signal Transduct. Target. Ther.* **2023**, *8*, 218. [CrossRef] [PubMed]

78. Alomri, R.M.; Kennedy, G.A.; Wali, S.O.; Alhejaili, F.; Robinson, S.R. Association between nocturnal activity of the sympathetic nervous system and cognitive dysfunction in obstructive sleep apnoea. *Sci. Rep.* **2021**, *11*, 11990. [CrossRef] [PubMed]

79. Liu, X.; Ma, Y.; Ouyang, R.; Zeng, Z.; Zhan, Z.; Lu, H.; Cui, Y.; Dai, Z.; Luo, L.; He, C.; et al. The relationship between inflammation and neurocognitive dysfunction in obstructive sleep apnea syndrome. *J. Neuroinflamm.* **2020**, *17*, 229. [CrossRef] [PubMed]

80. Ma, S.; Bi, W.; Liu, X.; Li, S.; Qiu, Y.; Huang, C.; Lv, R.; Yin, Q. Single-Cell Sequencing Analysis of the db/db Mouse Hippocampus Reveals Cell-Type-Specific Insights Into the Pathobiology of Diabetes-Associated Cognitive Dysfunction. *Front. Endocrinol.* **2022**, *13*, 891039. [CrossRef] [PubMed]

81. Xie, H.; Yung, W. Chronic intermittent hypoxia-induced deficits in synaptic plasticity and neurocognitive functions: A role for brain-derived neurotrophic factor. *Acta Pharmacol. Sin.* **2012**, *33*, 5–10. [CrossRef]

82. Ceccato, F.; Bernkopf, E.; Scaroni, C. Sleep apnea syndrome in endocrine clinics. *J. Endocrinol. Investig.* **2015**, *38*, 827–834. [CrossRef]

83. Li, M.; Li, X.; Lu, Y. Obstructive Sleep Apnea Syndrome and Metabolic Diseases. *Endocrinology* **2018**, *159*, 2670–2675. [CrossRef] [PubMed]

84. Katz, E.S.; D'Ambrosio, C.M. Pediatric Obstructive Sleep Apnea Syndrome. *Clin. Chest Med.* **2010**, *31*, 221–234. [CrossRef]

85. Drager, L.F.; Jun, J.C.; Polotsky, V.Y. Metabolic consequences of intermittent hypoxia: Relevance to obstructive sleep apnea. *Best Pract. Res. Clin. Endocrinol. Metab.* **2010**, *24*, 843–851. [CrossRef] [PubMed]

86. Söğüt, A.; Polotsky, V.Y. Leptin levels in children with obstructive sleep apnea syndrome. *Tuberk. Toraks* **2016**, *64*, 283–288. [CrossRef]

87. Drager, L.F.; Polotsky, V.Y. Lipid Metabolism: A New Frontier in Sleep Apnea Research. *Am. J. Respir. Crit. Care Med.* **2011**, *184*, 288–290. [CrossRef]

88. Chen, D.-D.; Huang, J.-F.; Lin, Q.-C.; Chen, G.-P.; Zhao, J.-M. Relationship between serum adiponectin and bone mineral density in male patients with obstructive sleep apnea syndrome. *Sleep Breath.* **2017**, *21*, 557–564. [CrossRef]

89. Justeau, G.; Gervès-Pinquié, C.; Le Vaillant, M.; Trzepizur, W.; Meslier, N.; Goupil, F.; Pigeanne, T.; Launois, S.; Leclair-Visonneau, L.; Masson, P.; et al. Association Between Nocturnal Hypoxemia and Cancer Incidence in Patients Investigated for OSA. *Chest* **2020**, *158*, 2610–2620. [CrossRef]

90. Brenner, R.; Kivity, S.; Peker, M.; Reinhorn, D.; Keinan-Boker, L.; Silverman, B.; Liphsitz, I.; Kolitz, T.; Levy, C.; Shlomi, D.; et al. Increased Risk for Cancer in Young Patients with Severe Obstructive Sleep Apnea. *Respiration* **2019**, *97*, 15–23. [CrossRef]

91. Nieto, F.J.; Peppard, P.E.; Young, T.; Finn, L.; Hla, K.M.; Farré, R. Sleep-disordered Breathing and Cancer Mortality. *Am. J. Respir. Crit. Care Med.* **2012**, *186*, 190–194. [CrossRef]

92. Ma, L.; Shan, W.; Ding, X.; Yang, P.; Rozjan, A.; Yao, Q. Intermittent hypoxia induces tumor immune escape in murine S180 solid tumors via the upregulation of TGF-β1 in mice. *Sleep Breath.* **2021**, *25*, 719–726. [CrossRef]

93. Nguyen, V.T.; Fields, C.J.; Ashley, N.T. Inflammation from Sleep Fragmentation Starts in the Periphery Rather than Brain in Male Mice. *Res. Sq.* **2023**, *13*, rs.3.rs-2544592.

94. Carreras, A.; Zhang, S.X.; Peris, E.; Qiao, Z.; Gileles-Hillel, A.; Li, R.C.; Wang, Y.; Gozal, D. Chronic sleep fragmentation induces endothelial dysfunction and structural vascular changes in mice. *Sleep* **2014**, *37*, 1817–1824. [CrossRef]

95. Salman, L.A.; Shulman, R.; Cohen, J.B. Obstructive Sleep Apnea, Hypertension, and Cardiovascular Risk: Epidemiology, Pathophysiology, and Management. *Curr. Cardiol. Rep.* **2020**, *22*, 6. [CrossRef]

96. Al-Halawani, M.; Kyung, C.; Liang, F.; Kaplan, I.; Moon, J.; Clerger, G.; Sabin, B.; Barnes, A.; Al-Ajam, M. Treatment of obstructive sleep apnea with CPAP improves chronic inflammation measured by neutrophil-to-lymphocyte ratio. *J. Clin. Sleep Med.* **2020**, *16*, 251–257. [CrossRef]

97. Ryan, S. Mechanisms of cardiovascular disease in obstructive sleep apnoea. *J. Thorac. Dis.* **2018**, *10*, S4201–S4211. [CrossRef]

98. Ohga, E.; Tomita, T.; Wada, H.; Yamamoto, H.; Nagase, T.; Ouchi, Y. Effects of obstructive sleep apnea on circulating ICAM-1, IL-8, and MCP-1. *J. Appl. Physiol.* **2003**, *94*, 179–184. [CrossRef]

99. Xu, H.; Wang, Y.; Guan, J.; Yi, H.; Yin, S. Effect of CPAP on Endothelial Function in Subjects With Obstructive Sleep Apnea: A Meta-Analysis. *Respir. Care* **2015**, *60*, 749–755. [CrossRef]

100. Fadaei, R.; Koushki, M.; Sharafkhaneh, A.; Moradi, N.; Ahmadi, R.; Rostampour, M.; Khazaie, H. The impact of continuous positive airway pressure therapy on circulating levels of malondialdehyde: A systematic review and meta-analysis. *Sleep Med.* **2020**, *75*, 27–36. [CrossRef]

101. Peker, Y.; Thunström, E.; Glantz, H.; Eulenburg, C. Effect of Obstructive Sleep Apnea and CPAP Treatment on Cardiovascular Outcomes in Acute Coronary Syndrome in the RICCADSA Trial. *J. Clin. Med.* **2020**, *9*, 4051. [CrossRef]

102. Nadeem, R.; Singh, M.; Nida, M.; Kwon, S.; Sajid, H.; Witkowski, J.; Pahomov, E.; Shah, K.; Park, W.; Champeau, D. Effect of CPAP treatment for obstructive sleep apnea hypopnea syndrome on lipid profile: A meta-regression analysis. *J. Clin. Sleep Med.* **2014**, *10*, 1295–1302. [CrossRef]

103. Iftikhar, I.H.; Hoyos, C.M.; Phillips, C.L.; Magalang, U.J. Meta-analyses of the Association of Sleep Apnea with Insulin Resistance, and the Effects of CPAP on HOMA-IR, Adiponectin, and Visceral Adipose Fat. *J. Clin. Sleep Med.* **2015**, *11*, 475–485. [CrossRef]

104. Poyares, D.; Oliveira, W.; Fujita, L. Can CPAP prevent myocardial damage? *Anatol. J. Cardiol.* **2014**, *14*, 272–273. [CrossRef]

105. Berbenetz, N.; Wang, Y.; Brown, J.; Godfrey, C.; Ahmad, M.; Vital, F.M.; Lambiase, P.; Banerjee, A.; Bakhai, A.; Chong, M. Non-invasive positive pressure ventilation (CPAP or bilevel NPPV) for cardiogenic pulmonary oedema. *Cochrane Database Syst. Rev.* **2019**, *4*, CD005351. [CrossRef]

106. Chasens, E.R.; Korytkowski, M.; Burke, L.E.; Strollo, P.J.; Stansbury, R.; Bizhanova, Z.; Atwood, C.W.; Sereika, S.M. Effect of Treatment of OSA With CPAP on Glycemic Control in Adults With Type 2 Diabetes: The Diabetes Sleep Treatment Trial (DSTT). *Endocr. Pract.* **2022**, *28*, 364–371. [CrossRef]

107. Pollicina, I.; Maniaci, A.; Lechien, J.R.; Iannella, G.; Vicini, C.; Cammaroto, G.; Cannavicci, A.; Magliulo, G.; Pace, A.; Cocuzza, S.; et al. Neurocognitive Performance Improvement after Obstructive Sleep Apnea Treatment: State of the Art. *Behav. Sci.* **2021**, *11*, 180. [CrossRef]

108. Uniken Venema, J.A.M.; Rosenmöller, B.R.A.M.; de Vries, N.; de Lange, J.; Aarab, G.; Lobbezoo, F.; Hoekema, A. Mandibular advancement device design: A systematic review on outcomes in obstructive sleep apnea treatment. *Sleep Med. Rev.* **2021**, *60*, 101557. [CrossRef]

109. Lee, W.; Nagubadi, S.; Kryger, M.H.; Mokhlesi, B. Epidemiology of Obstructive Sleep Apnea: A Population-based Perspective. *Expert. Rev. Respir. Med.* **2008**, *2*, 349–364. [CrossRef]

110. Ming, H.; Tian, A.; Liu, B.; Hu, Y.; Liu, C.; Chen, R.; Cheng, L. Inflammatory cytokines tumor necrosis factor-α, interleukin-8 and sleep monitoring in patients with obstructive sleep apnea syndrome. *Exp. Ther. Med.* **2019**, *17*, 1766–1770. [CrossRef]

111. Tauman, R.; O'Brien, L.M.; Gozal, D. Hypoxemia and obesity modulate plasma C-reactive protein and interleukin-6 levels in sleep-disordered breathing. *Sleep Breath.* **2007**, *11*, 77–84. [CrossRef]

112. Galic, T.; Bozic, J.; Ivkovic, N.; Gunjaca, G.; Ticinovic, T.K.; Dogas, Z. Effects of mandibular advancement device treatment on arterial stiffness and glucose metabolism in patients with mild to moderate obstructive sleep apnea: A prospective 1 year study. *Sleep Breath.* **2016**, *20*, 69–77. [CrossRef]

113. Recoquillon, S.; Pépin, J.L.; Vielle, B.; Andriantsitohaina, R.; Bironneau, V.; Chouet-Girard, F.; Fleury, B.; Goupil, F.; Launois, S.; Martinez, M.C.; et al. Effect of mandibular advancement therapy on inflammatory and metabolic biomarkers in patients with severe obstructive sleep apnoea: A randomised controlled trial. *Thorax* **2019**, *74*, 496–499. [CrossRef]

114. Trzepizur, W.; Gagnadoux, F.; Abraham, P.; Rousseau, P.; Meslier, N.; Saumet, J.L.; Racineux, J.L. Microvascular endothelial function in obstructive sleep apnea: Impact of continuous positive airway pressure and mandibular advancement. *Sleep Med.* **2009**, *10*, 746–752. [CrossRef]

115. Lee, L.A.; Huang, C.G.; Chen, N.H.; Wang, C.L.; Fang, T.J.; Li, H.Y. Severity of obstructive sleep apnea syndrome and high-sensitivity C-reactive protein reduced after relocation pharyngoplasty. *Otolaryngol. Head Neck Surg.* **2011**, *144*, 632–638. [CrossRef]

116. Binar, M.; Akcam, T.; Karako, O.; Sagkan, R.I.; Musabak, U.; Gerek, M. A new surgical technique versus an old marker: Can expansion sphincter pharyngoplasty reduce C-reactive protein levels in patients with obstructive sleep apnea? *Eur. Arch. Otorhinolaryngol.* **2017**, *274*, 829–836. [CrossRef]

117. Iannella, G.; Lechien, J.R.; Perrone, T.; Meccariello, G.; Cammaroto, G.; Cannavicci, A.; Burgio, L.; Maniaci, A.; Cocuzza, S.; Di Luca, M.; et al. Barbed reposition pharyngoplasty (BRP) in obstructive sleep apnea treatment: State of the art. *Am. J. Otolaryngol.* **2022**, *43*, 103197. [CrossRef]

Brief Report

Sleep Architecture and Daytime Sleepiness in Patients with Erectile Dysfunction

Helena Martynowicz [1], Rafal Poreba [1], Tomasz Wieczorek [2], Zygmunt Domagala [3], Robert Skomro [4], Anna Wojakowska [1], Sylwia Winiewska [5], Piotr Macek [1], Grzegorz Mazur [1] and Paweł Gac [6,*]

[1] Department of Internal Medicine, Occupational Diseases, Hypertension and Clinical Oncology, Wroclaw Medical University, Borowska 213, 50-556 Wroclaw, Poland
[2] Department of Psychiatry, Wroclaw Medical University, Pasteura 10, 50-367 Wroclaw, Poland
[3] Department of Anatomy, Wroclaw Medical University, Chałubińskiego 6a, 50-368 Wroclaw, Poland
[4] Division of Respiratory Critical Care and Sleep Medicine, University of Saskatchewan, Saskatoon, SK S7N 5E5, Canada
[5] Department of Experimental Dentistry, Wroclaw Medical University, Krakowska 26, 50-425 Wroclaw, Poland
[6] Division of Environmental Health and Occupational Medicine, Department of Population Health, Wroclaw Medical University, Mikulicza-Radeckiego 7, 50-368 Wroclaw, Poland
* Correspondence: pawelgac@interia.pl or pawel.gac@umw.edu.pl

Abstract: Obstructive sleep apnea is considered a risk factor for erectile dysfunction. The aim of this study was to determine sleep architecture and assess daytime sleepiness in patients with erectile dysfunction. The study group included 280 patients. The 107 enrolled patients had reported erectile dysfunction. The control group consisted of 173 patients who had no history of erectile dysfunction. The Epworth sleepiness scale (ESS) was used to measure the subjects' level of daytime sleepiness. All patients underwent a standardized overnight, single-night polysomnography in sleep laboratory. In the erectile dysfunction group, we observed increased ESS total score and N1 sleep phase duration. Mean and minimal oxygen saturation and mean oxygen desaturation were decreased in comparison to the control group. In summary, subjects with erectile dysfunction have altered sleep architecture, oxygen saturation parameters and increased daytime sleepiness.

Keywords: hypertension; excessive daytime sleepiness; sleep apnea; sexual dysfunction

check for **updates**

Citation: Martynowicz, H.; Poreba, R.; Wieczorek, T.; Domagala, Z.; Skomro, R.; Wojakowska, A.; Winiewska, S.; Macek, P.; Mazur, G.; Gac, P. Sleep Architecture and Daytime Sleepiness in Patients with Erectile Dysfunction. *Life* **2023**, *13*, 1541. https://doi.org/10.3390/life13071541

Academic Editor: Konstantinos Chaidas

Received: 25 May 2023
Revised: 5 July 2023
Accepted: 10 July 2023
Published: 11 July 2023

1. Introduction

Erectile dysfunction (ED) is defined as the consistent inability to obtain and/or maintain a penile erection during sexual activity [1]. ED affects millions of middle-aged to elderly men worldwide [2]. There are many causes of ED including diabetes, ischemic heart disease, medications (e.g., thiazides, b-blockers, spironolactone, and antidepressants), neurogenic disorders, atherosclerosis, tobacco use, hyperlipidemia, hypogonadism, lower urinary tract symptoms, metabolic syndrome, and depression [3,4]. The prevalence of ED increases with age, particularly after the age of 60 years [4–6]. There also data suggesting that obstructive sleep apnea (OSA) may have an independent association with sexual dysfunction and impotence [7,8]. OSA is a common sleep disorder characterized by the collapse of the upper airway leading to the cessation of airflow, intermittent arterial oxygen desaturation and arousals during sleep. Recent evidence showed that one in five adults suffer from at least a mild degree of OSA, 936 million adults aged 30–69 years have mild-to-severe OSA and 425 million (399–450) adults aged 30–69 years have moderate-to-severe obstructive sleep apnea globally [9,10]. Thus, this is one of the most common sleep disorders. Male sex and obesity are known risk factors for sleep apnea [11]. Several studies confirmed the increased prevalence of ED in patients with OSA [12,13].

Overnight polysomnography (PSG) is a gold standard in diagnosis of OSA [14]. Due to limited accessibility of PSG and its high expenditure, alternative tools for screening purposes have been developed including the Epworth sleepiness scale, Berlin Questionnaire or STOB-BANG Questionnaire. One of the most widely used is the Epworth sleepiness scale (ESS), which measures general level of sleepiness. The scale is self-administered: patients estimate their probability of falling asleep during different situations. The tool has been used in normal subjects [15], as well as in those with OSA [16], narcolepsy [17], stroke [18], coronary artery disease [19], heart failure [20], epilepsy [21], Parkinson's disease [22], hemodialysis [23], diabetes [24], rheumatoid arthritis [25] and obesity [24]. However, the data concerning daytime sleepiness in erectile dysfunctions are limited.

The aims of this study were as follows: 1. to determine sleep architecture in ED patients; and to 2. assess sleepiness scores using the Epworth sleepiness scale (ESS) in ED patients.

2. Materials and Methods

A summary of the study protocol is shown in Figure 1.

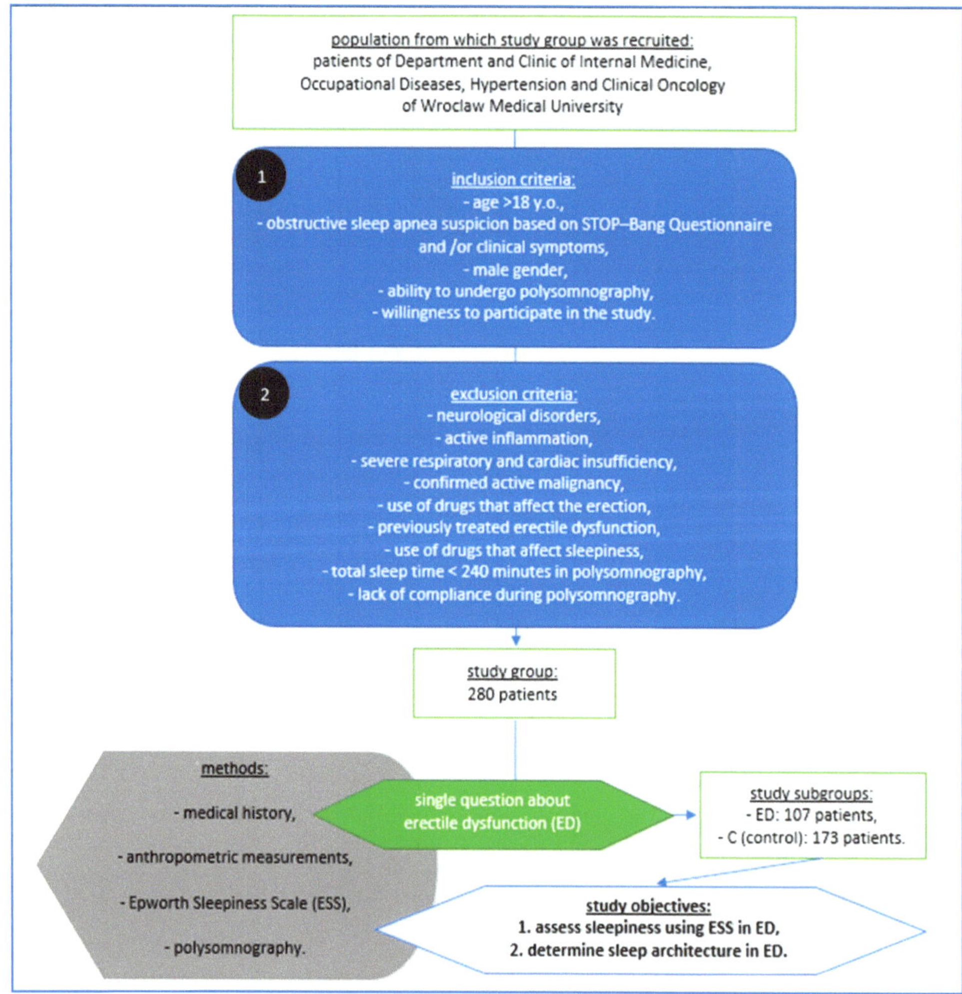

Figure 1. Material, methods and objectives of the study according to its protocol.

The study group included 280 male patients of Department and Clinic of Internal Medicine, Occupational Diseases, Hypertension and Clinical Oncology hospitalized for the assessment of possible obstructive sleep apnea. The inclusion criteria obtain age > 18 years old, obstructive sleep apnea suspicion based on STOP–Bang Questionnaire and/or clinical symptoms, male gender, ability to undergo polysomnography and willingness to participate in the study, while the exclusion criteria included the presence of neurological disorders, active inflammation, severe respiratory and cardiac insufficiency, confirmed active malignancy, the use of drugs that affect the erection, previously treated erectile dysfunction, the use of drugs that affect sleepiness, in polysomnography total sleep time < 240 min and a lack of compliance during the study. The 107 enrolled patients had reported erectile dysfunction (ED). The control group (C) consisted of 173 patients who had no history of erectile dysfunction. ED was assessed via a single question during a clinical interview [26,27]. The Epworth sleepiness scale (ESS) was used to measure the subjects' level of daytime sleepiness. The ESS was developed by Murray Johns at Epworth Hospital in Australia and was first reported in 1991. In the Epworth scale, the patients rate dozing in eight different situations. The minimum score of 0 indicates "would never doze", while a maximum score of 3 indicates "high chance of dozing". The total score can range from a minimum of 0 to a maximum of 24. Scores $\geq$ 10 on the ESS were indicative of excessive daytime sleepiness [15,16].

Height and weight were recorded using a nursing calibrated scale. The body mass index (BMI, calculated as weight in kilogram divided by square of height in meter) was calculated. Besides ESS, a questionnaire on OSA symptoms, OSA comorbidities, smoking status was performed. In the study group, 37.50% were smokers ($n = 105$), 78.21% hypertensives ($n = 219$), 12.85% patients had coronary heart disease ($n = 36$), 6.7% patients had a history of myocardial infarction ($n = 19$), and 8.2% were assessed after a stroke ($n = 23$). The mean score of the Epworth scale was 9.33 $\pm$ 5.40. The mean PSG parameters of the entire study group are presented in Table 1.

Table 1. The PSG parameters of the entire study group.

PSG Parameter	Mean Value	SD
TST (min)	384.53	108.77
Sleep latency (min)	31.0	31.87
SE (%)	73.82	18.82
AHI (n/h)	21.83	20.63
Mean O_2 Sat (%)	93.71	2.84
ODI (min/h)	20.44	21.17
Mean desaturation (%)	4.95	1.81
Min O_2 Sat(O_2)	81.16	8.80
SatO_2 < 90% (%)	14.85	22.21
N1 (%TST)	11.73	14.14
N2 (%TST)	49.94	18.93
N3 (%TST)	19.56	17.73
REM (%TST)	19.21	10.18

PSG—polysomnography; TST—total sleep time; AHI—apnea/hypopnea index; ODI—oxygen desaturation index; mean SpO$_2$—mean oxygen saturation; min SpO$_2$—minimal oxygen saturation; REM—rapid eye movements.

All patients underwent a standardized overnight, single-night polysomnography in a sleep laboratory. We used the NOXA1 (NOX Medical) PSG system. Polysomnograms were assessed in 30 s epochs according to the AASM (American Academy of Sleep Medicine) standard criteria for sleep scoring. PSG outcome variables included sleep latency, total sleep time (TST) and sleep efficiency (%), the ratio of N1, N2, N3 and the stage of REM. Abnormal respiratory events were scored from the pressure airflow signal evaluated according to the standard criteria of the American Academy of Sleep Medicine Task Force [28]. Apneas were defined as the absence of airflow for $\geq$10 s. Hypopnea was defined as a reduction in the amplitude of breathing by $\geq$30% for $\geq$10 s with a decline $\geq$3% in blood oxygen saturation or an arousal. A NONIN WristOx2 3150 pulse oximeter (Nonin Medical Inc., Plymouth, MN, USA), coupled with the PSG system, was used to record the oxygen saturation level. To analyze the full polysomnography recording, the Noxturnal software (Nox Medical, Reykjavík, Iceland) was used. A certified, qualified physician (H.M.) from the sleep laboratory scored and manually analyzed the data in accordance with the AASM guidelines.

Statistical analysis was conducted using the statistical software Statistica 12 PL, Statistica, Tulsa, US. For quantitative variables, arithmetic means and standard deviations were calculated for the estimated parameters in the studied groups. The distribution of the variables was tested using the Shapiro–Wilk test. In cases of quantitative variables manifesting the normal distribution in further statistical analysis, the t test for unlinked variables was used. In cases of variables manifesting distribution distinct from the normal one, the nonparametric equivalent of the t test, i.e., the Mann–Whitney U test was used. In order to detect relationships between the studied variables, univariate regression analysis was performed. The results at the level of $p < 0.05$ were accepted as statistically significant.

This study was approved by the Ethical Committee of Wroclaw Medical University (ID KB-227/2015) and was conducted in accordance with the Declaration of Helsinki. All patients signed an informed consent form for this study.

3. Results

The mean BMI was higher in the ED group compared to the that of the control. The mean age, BMI, height, body mass and smoking status are shown in Table 2.

Table 2. The height, body mass, BMI (body mass index), and smoking status of patients with erectile dysfunction (ED) and control (C) subjects.

	C (n = 173)	ED (n = 107)	p
Age (years)	48.49 ± 13.31	48.92 ± 10.30	ns
Height (cm)	177.13 ± 7.44	174.76 ± 6.84	<0.05
Body mass (kg)	95.83 ± 18.90	99.57 ± 17.49	ns
BMI (kg/m^2)	30.50 ± 5.46	32.52 ± 4.73	<0.01
Smoking (years)	20.45 ± 11.70	23.55 ± 12.61	ns

In the ED group, statistically significantly AHI (apnea/hypopnea index) < 5 (AHI excluding OSA) was observed less frequently than in the control group. The number of subjects according to AHI in patients with erectile dysfunction (ED) and control (C) subjects is shown in Table 3.

Table 3. The number of subjects according to AHI in patients with erectile dysfunction (ED) and control (C) subjects.

AHI	C (n = 173)	ED (n = 107)	p
<5 (n = 44)	32 (24)	12 (13)	<0.05
>5 and <15 (n = 66)	39 (30)	27 (30)	ns
>15 and <30 (n = 51)	27 (21)	24 (26)	ns
>30 (n = 61)	33 (25)	28 (31)	ns

AHI—apnea/hypopnea index.

In the erectile dysfunction group (ED), we observed a higher ESS total score and lower mean and minimal oxygen saturation levels compared to those of the control group. The mean oxygen desaturation levels were lower in the ED group in comparison with the control group. The N1 sleep phase was increased in comparison with the control group. The polysomnographic parameters and ESS (Epworth sleepiness scale) score in patients with erectile dysfunction (ED) and control (C) subjects are presented in Table 4.

Table 4. The polysomnographic parameters and ESS (Epworth sleepiness scale) score in patients with erectile dysfunction (ED) and control (C) subjects.

	C (*n* = 173)	ED (*n* = 107)	*p*
ESS score (points)	8.87 ± 5.33	10.07 ± 5.44	<0.05
TST (min)	399.92 ± 110.83	360.55 ± 102.35	ns
Sleep latency (min)	29.38 ± 31.26	33.75 ± 33.13	ns
AHI (/h)	20.41 ± 20.65	23.87 ± 20.54	ns
Mean SpO_2(%)	94.19 ± 2.12	92.98 ± 3.57	<0.01
Min SpO_2(%)	82.24 ± 2.11	79.56 ± 9.68	<0.05
ODI (/h)	18.85 ± 20.63	22.84 ± 21.87	ns
Mean oxygen desaturation (%)	4.67 ± 1.79	5.42 ± 1.78	<0.05
Mean HR (bpm)	60.72 ± 8.62	62.09 ± 9.51	ns
N1 (%TST)	8.35 ± 6.93	17.34 ± 20.22	<0.01
N2 (%TST)	52.32 ± 19.24	46.01 ± 18.00	ns
N3 (%TST)	19.75 ± 17.87	19.26 ± 17.78	ns
REM (%TST)	20.35 ± 10.06	17.39 ± 10.26	ns
Sleep efficiency (%)	76.30 ± 17.78	70.06 ± 19.69	ns

TST—total sleep time; AHI—apnea/hypopnea index; ODI—oxygen desaturation index; Mean SpO_2—mean oxygen saturation; Min SpO_2—minimal oxygen saturation; REM—rapid eye movements.

Then, we performed an additional analysis considering the occurrence of hypertension. Hypertension is an important risk factor for erectile dysfunction and may also affect the sleep architecture and the level of sleepiness. Therefore, it might affect the results of the study. We divided all participants into two subgroups (normotensive and hypertensives). In univariate regression analysis, in the hypertensive group, we observed a relationship between the presence of erectile dysfunction and mean saturation (r = −0.20; $p < 0.05$), minimal saturation (r = −0.14; $p < 0.05$), and N1 (%TST) (r = 0.31, $p < 0.05$). In the normotensive group, we observed a relationship between ED and AHI (apnea/hypopnea index) (r = 0.43, $p < 0.05$), ED and ODI (r = 0.45, $p < 0.05$) and ED and TST (r = −0.82, $p < 0.05$). In this group, we also observed a relationship between ED and mean saturation (r = −0.44, $p < 0.05$) and minimal saturation (r = −0.45, $p < 0.05$). Interestingly, in the normotensive group, we observed a relationship between ED and N2 (%TST) (r = −0.57, $p < 0.05$), N3 (%TST) (r = 0.62, $p < 0.05$) and REM (rapid eye movement) sleep (%TST) (r = −0.61, $p < 0.05$). The result of the univariate regression analyses in the hypertensive and normotensive groups are summarized in Table 5.

Table 5. The results of univariate regression analyzes between erectile dysfunction (ED) and polysomnographic parameters in hypertensive and normotensive groups.

Dependent Variable	Independent Variable	Hypertensive Group	Normotensive Group
Presence of erectile dysfunction	ESS score (points)	ns	ns
	TST (min)	ns	r = −0.82
	Sleep latency (min)	ns	ns
	AHI (/h)	ns	r = 0.43
	Mean SpO_2 (%)	r = −0.20	r = −0.44
	Min SpO_2 (%)	r = −0.14	r = −0.45
	ODI (/h)	ns	r = 0.45
	Mean oxygen desaturation (%)	ns	ns
	Mean HR (bpm)	ns	ns
	N1 (%TST)	r = 0.31	ns
	N2 (%TST)	ns	r = −0.57
	N3 (%TST)	ns	r = 0.63
	REM (%TST)	ns	r = −0.61
	Sleep efficiency (%)	ns	ns

TST—total sleep time; AHI—apnea/hypopnea index; ODI—oxygen desaturation index; Mean SpO_2—mean oxygen saturation; Min SpO_2—minimal oxygen saturation; REM—rapid eye movements.

4. Discussion

Obstructive sleep apnea affects both the patient and the sexual partner [29–31]. In a cross-sectional study of 401 men referred to a sleep lab for suspected OSA, 92% were diagnosed with OSA, and ED was present in 69% of those with OSA and 34% without OSA [8]. The evidence for the association between OSA and ED remains inconclusive, with some studies suggesting no association, or an association being shown for severe OSA only [32,33]. The pathogenesis of ED in OSA is complex. The factors implicated include increased sympathetic activity, oxidative stress, endothelial dysfunction a reduction in nitric oxide formation [34,35]. Sleep fragmentation and a lack of rapid eye movement (REM) periods related to OSA may be involved in ED pathogenesis because physiological erections appear to help maintain erectile function through cavernous tissue oxygenation during the REM sleep period [36,37]. ED is also related to changes in the hormonal axis caused by sleep pattern changes associated with OSA [38].

The main result of this study is an increased ESS total score in patients with erectile dysfunction compared to those of the control group, although the AHI and ODI (oxygen desaturation index) were similar in both groups. It is worth noting that this is a novel observation. In the ED group, the mean saturation, minimal saturation and mean desaturation levels were decreased compared to those of control group. Interestingly, we also observed increased N1 sleep duration (%TST) in the ED group. N1 is the lighter stage of NREM sleep, which usually occurs at the beginning of sleep and often alternates with brief arousal episodes. N1 is the period of transition from unsynchronized beta and gamma brain waves to more synchronized but slower alpha waves, and then to theta waves with slow rolling eye movement. This stage comprises only about 5% of the total sleep time. In our study, N1 comprises as much as 17.3% of TST in patients with erectile dysfunction. This phenomenon may be a consequence of increased activity of the sympathetic nervous system, which is common in OSA due to repeated hypoxemia. Additionally, fatigue and a decrease in rapid eye movement (REM) sleep period may provoke the deterioration in the quality of erections [39,40]. These results agree with those of our study. We observed a negative relationship between ED and the REM (rapid eye movement) period (%TST); however, we did not observe statistical differences between the REM period in ED and controls.

In our study, the patients with ED had lower mean oxygen saturation, minimal saturation and mean desaturation levels compared to the control group. These results are in agreement with those of previous studies, which showed that recurrent apnea attacks in patients with OSA cause hypoxia reperfusion injury and oxidative stress, endothelial dysfunction and, consequently, ED [41]. Severe and moderate OSA was more prevalent in the ED group than in the control group, which may be a possible explanation for these observations.

OSA is a known risk factor for erectile dysfunction; however, the mechanism underlying ED in patients with OSA is complex and remains unclear. The data on EDS (excessive daytime sleepiness) and ED are limited and often contradictory. Surprisingly, it was demonstrated that ED subjects had significantly lower ESS and SaO_2 [42]. Recently, Jeon also showed that ESS is inversely correlated with the International Index of Erectile Function (KIIEF-5) [43]. We have shown that patients with ED have increased total ESS scores compared to those of the control group, which is a novel observation. This is most likely the result of increased frequency of severe and moderate OSA in the ED group. However, we have not observed a statistically significant difference in AHI between the ED and control groups; thus, other mechanisms may be involved.

The most unexpected result of our study is the presence of a relationship between ED and N3 sleep (%TST) in normotensive patients; however, the duration of N3 was similar between the ED group and the control group. We did not observe this relationship in hypertensive patients, which might suggest different relationships between the structure of sleep and ED depending on the blood pressure levels. We have also observed a positive relationship between AHI, ODI (oxygen desaturation index), and erectile dysfunction, and a negative relationship between mean saturation, minimal saturation, TST and erectile dysfunction in the normotensive group. The relationship between AHI, ODI, TST and ED was not observed in hypertensive patients.

The present study confirms the strong role of hypoxia in erectile dysfunction. Firstly, we observed decreased mean oxygen saturation as well as minimal oxygen saturation levels in the erectile dysfunction group compared to those of the control group. Secondly, both in hypertensive and normotensive patients, mean and minimal oxygen saturation were related to erectile dysfunction. Chronic hypoxia may be observed in physiological conditions such as aging, as well as in many pathological conditions such as smoking and chronic obstructive pulmonary disease (COPD), heart and respiratory failure, obstructive sleep apnea, atherosclerosis, diabetes, and hypertension. It is worth noting that CPAP (continuous positive airway pressure), when used with testosterone replacement therapy, can increase total testosterone and cause an improvement in the indicators of altered nocturnal penile erection episodes [44]. In the present study's regression analysis, the oxygen desaturation index was associated with ED only in the normotensive group. These results may indicate the role of oxygen level drops in normotensive but not in hypertensive patients. Recently, Feng demonstrated that minimum oxygen saturation and average oxygen saturation may predict the occurrence of ED in obstructive sleep apnea patients [45]. The possible mechanisms involved in hypoxia on erectile dysfunction mainly include excessive reactive-oxygen-species-mediated oxidative stress, hypoxia-inducible factors-1α mediated endothelial cell apoptosis and proliferation inhibition, endothelial dysfunction, reduced nitric oxide production and systematic inflammation [46].

Short sleep duration is considered a risk factor for erectile dysfunction. The sleep duration recommended by the AASM is between 7 and 9 h of sleep. A short sleep duration is defined as habitual sleep time of less than 6 h [47]. It was previously shown that short sleep is associated with a large group of disorders, such as hypertension, diabetes, major depressive disorder, and other morbidities [48,49]. Sleep duration is decreasing in modern societies, as large cohorts studies have shown [50–52]. As much as 29.1% of adults suffered from short sleep duration is United States [53]. The pathomechanisms of erectile dysfunction in short sleepers are complex and lead to hypothalamic–pituitary–adrenal axis overactivity, resulting in cortisol release as well as an autonomic nervous system imbalance, consequently resulting in catecholamines release and vasoconstriction. Moreover, short sleep duration may reduce the testosterone level and the frequency of REM sleep and sleep-related erections (SREs) [54]. SREs or nocturnal penile tumescence (NPT) can be measured using PSG. The majority of NPT occurrences, which are a physiological and spontaneous phenomenon, are related to REM sleep [55]. The erection initiates during the shift from NREM to REM sleep with full tumescence throughout REM; however, in the present study, we did not measure NPT. It is worth noting that short sleep duration is related to many negative health outcomes such as cardiovascular diseases, increased morbidity and mortality [56,57]. In the present study, we did not observe statistically significant differences in total sleep time between the patients with erectile dysfunction and the control group. Although sleep time was comparable, we noticed an increased stage N1 sleep duration. A high percentage of the stage N1 sleep is usually a result of frequent arousals caused by sleep disorders or environmental disturbances. Thus, in the present study, altered sleep structure, but not sleep duration, was related to erectile dysfunction.

Summarizing, we studied the sleep structure and level of sleepiness of a relatively large ED cohort using polysomnography, which is the gold standard for sleep assessment. We have confirmed many observations that have been described before, especially the relationship between OSA and ED; however, this study has some novel conclusions. We have also observed an increased sleepiness level in ED subjects in comparison to that of the control group. We have shown the importance of oxygen saturation parameters and altered sleep architecture in erectile dysfunction. We have also confirmed the relationship between ED and sleep parameters in normotensive but not in hypertensive patients. These results indicate a different mechanism of ED in hypertension; thus, further studies are needed to explain these observations.

The strengths of this study are its large population and use of the gold standard in sleep disorders diagnosis (avPSG). However, several limitations of the present study should be highlighted. Firstly, we did not use any scales for measuring erectile dysfunction, including the International Index of Erectile Dysfunction (IIED), which may be a major confounder. The cause, type and degree of severity of ED was not studied. We did not collect data on hypotensive therapy, which could affect erectile dysfunction. In addition, there was a small number of patients in the group without OSA compared to the OSA group.

5. Conclusions

Patients with erectile dysfunction are more likely to be affected by obstructive sleep apnea, altered sleep architecture and oxygen saturation parameters compared to the control group. The sleepiness measured using ESS in patients with erectile dysfunction is increased compared to that of the control group.

Author Contributions: Conceptualization, H.M.; Investigation, H.M., T.W., Z.D., A.W., S.W. and P.M.; Methodology, H.M.; Project administration, H.M.; Software, R.P. and P.G.; Supervision, H.M., R.P., R.S., G.M. and P.G.; Writing—original draft, H.M.; Writing—review and editing, P.G. All authors have read and agreed to the published version of the manuscript.

Funding: This work was supported by the Wroclaw Medical University (project number as recorded in the Simple system SUBZ.E264.23.039).

Institutional Review Board Statement: The study was conducted according to the guidelines of the Declaration of Helsinki, and approved by the Ethics Committee of Wroclaw Medical University.

Informed Consent Statement: Informed consent was obtained from all subjects involved in the study.

Data Availability Statement: The data presented in this study are available upon request from the corresponding author. The data are not publicly available.

Conflicts of Interest: The authors declare no conflict of interest.

References

1. Rosen, R.C.; Riley, A.; Wagner, G.; Osterloh, I.H.; Kirkpatrick, J.; Mishra, A. The International Index of Erectile Function (IIEF): A multidimensional scale for assessment of erectile dysfunction. *Urology* **1997**, *49*, 822–830. [CrossRef] [PubMed]
2. Grant, P.; Jackson, G.; Baig, I.; Quin, J. Erectile dysfunction in general medicine. *Clin. Med.* **2013**, *13*, 136–140. [CrossRef] [PubMed]
3. Kałka, D.; Gebala, J.; Borecki, M.; Pilecki, W.; Rusiecki, L. Return to sexual activity after myocardial infarction—An analysis of the level of knowledge in men undergoing cardiac rehabilitation. *Eur. J. Intern. Med.* **2017**, *37*, e31–e33. [CrossRef]
4. Shamloul, R.; Ghanem, H. Erectile dysfunction. *Lancet* **2013**, *381*, 153–165. [CrossRef]
5. McVary, K.T. Clinical practice. Erectile dysfunction. *N. Engl. J. Med.* **2007**, *357*, 2472–2481. [CrossRef] [PubMed]
6. Johannes, C.B.; Araujo, A.B.; Feldman, H.A.; Derby, C.A.; Kleinman, K.P.; McKinlay, J.B. Incidence of erectile dysfunction in men 40 to 69 years old: Longitudinal results from the Massachusetts male aging study. *J. Urol.* **2000**, *163*, 460–463. [CrossRef] [PubMed]
7. Stannek, T.; Hürny, C.; Schoch, O.D.; Bucher, T.; Münzer, T. Factors affecting self-reported sexuality in men with obstructive sleep apnea syndrome. *J. Sex Med.* **2009**, *6*, 3415–3424. [CrossRef]
8. Budweiser, S.; Enderlein, S.; Jörres, R.A.; Hitzl, A.P.; Wieland, W.F.; Pfeifer, M.; Arzt, M. Sleep apnea is an independent correlate of erectile and sexual dysfunction. *J. Sex Med.* **2009**, *6*, 3147–3157. [CrossRef]
9. Young, T.; Peppard, P.E.; Gottlieb, D.J. Epidemiology of obstructive sleep apnea: A population health perspective. *Am. J. Respir. Crit Care Med.* **2002**, *165*, 1217–1239. [CrossRef]
10. Benjafield, A.V.; Ayas, N.T.; Eastwood, P.R.; Heinzer, R.; Ip, M.S.M.; Morrell, M.J.; Nunez, C.M.; Patel, S.R.; Penzel, T.; Pépin, J.L.; et al. Estimation of the global prevalence and burden of obstructive sleep apnoea: A literature-based analysis. *Lancet Respir. Med.* **2019**, *7*, 687–698. [CrossRef]
11. Young, T.; Palta, M.; Dempsey, J.; Skatrud, J.; Weber, S.; Badr, S. The occurrence of sleep-disordered breathing among middle-aged adults. *N. Engl. J. Med.* **1993**, *328*, 1230–1235. [CrossRef] [PubMed]
12. Margel, D.; Cohen, M.; Livne, P.M.; Pillar, G. Severe, but not mild, obstructive sleep apnea syndrome is associated with erectile dysfunction. *Urology* **2004**, *63*, 545–549. [CrossRef] [PubMed]

13. Perimenis, P.; Konstantinopoulos, A.; Karkoulias, K.; Markou, S.; Perimeni, P.; Spyropoulos, K. Sildenafil combined with continuous positive airway pressure for treatment of erectile dysfunction in men with obstructive sleep apnea. *Int. Urol. Nephrol.* **2007**, *39*, 547–552. [CrossRef]

14. Malhotra, A.; White, D.P. Obstructive sleep apnoea. *Lancet* **2002**, *360*, 237–245. [CrossRef] [PubMed]

15. Johns, M.W. A new method for measuring daytime sleepiness—The epworth sleepiness scale. *Sleep* **1991**, *14*, 540–545. [CrossRef]

16. Fong, S.Y.; Ho, C.K.; Wing, Y.K. Comparing MSLT and ESS in the measurement of excessive daytime sleepiness in obstructive sleep apnoea syndrome. *J. Psychosom. Res.* **2005**, *58*, 55–60. [CrossRef] [PubMed]

17. Van der Heide, A.; van Schie, M.K.; Lammers, G.J.; Dauvilliers, Y.; Arnulf, I.; Mayer, G.; Bassetti, C.; Ding, C.-L.; Lehert, P.; Van Dijk, J.G. Comparing Treatment Effect Measurements in Narcolepsy: The Sustained Attention to Response Task, Epworth Sleepiness Scale and Maintenance of Wakefulness Test. *Sleep* **2015**, *38*, 1051–1058. [CrossRef]

18. Mills, R.J.; Koufali, M.; Sharma, A.; Tennant, A.; Young, C.A. Is the Epworth sleepiness scale suitable for use in stroke? *Top. Stroke Rehabil.* **2013**, *20*, 493–499. [CrossRef]

19. Lee, C.H.; Ng, W.Y.; Hau, W.; Ho, H.H.; Tai, B.C.; Chan, M.Y.; Richards, A.M.; Tan, H.-C. Excessive daytime sleepiness is associated with longer culprit lesion and adverse outcomes in patients with coronary artery disease. *Sleep Med.* **2013**, *15*, 1267–1272. [CrossRef]

20. Wu, X.; Fu, C.; Zhang, S.; Liu, Z.; Li, S.; Jiang, L. Adaptive servoventilation improves cardiac dysfunction and prognosis in heart failure patients with sleep-disordered breathing: A meta-analysis. *Clin. Respir. J.* **2015**, *25*, 547–557. [CrossRef]

21. Quigg, M.; Gharai, S.; Ruland, J.; Schroeder, C.; Hodges, M.; Ingersoll, K.S.; Thorndike, F.P.; Yan, G.; Ritterband, L.M. Insomnia in epilepsy is associated with continuing seizures and worse quality of life. *Epilepsy Res.* **2016**, *122*, 91–96. [CrossRef]

22. Cochen De Cock, V.; Bayard, S.; Jaussent, I.; Charif, M.; Grini, M.; Langenier, M.C.; Yu, H.; Lopez, R.; Geny, C.; Carlander, B.; et al. Daytime sleepiness in Parkinson's disease: A reappraisal. *PLoS ONE* **2014**, *9*, e107278. [CrossRef] [PubMed]

23. Shen, Q.; Huang, X.; Luo, Z.; Xu, X.; Zhao, X.; He, Q. Sleep quality, daytime sleepiness and health-related quality-of-life in maintenance haemodialysis patients. *J. Int. Med. Res.* **2016**, *44*, 698–709. [CrossRef] [PubMed]

24. Gomez-Peralta, F.; Abreu, C.; Castro, J.C.; Alcarria, E.; Cruz-Bravo, M.; Garcia-Llorente, M.J.; Albornos, C.; Moreno, C.; Cepeda, M.; Almódovar, F. An association between liraglutide treatment and reduction in excessive daytime sleepiness in obese subjects with type 2 diabetes. *BMC Endocr. Disord.* **2015**, *4*, 78. [CrossRef] [PubMed]

25. Purabdollah, M.; Lakdizaji, S.; Rahmani, A.; Hajalilu, M.; Ansarin, K. Relationship between Sleep Disorders, Pain and Quality of Life in Patients with Rheumatoid Arthritis. *J. Caring Sci.* **2015**, *4*, 233–241. [CrossRef]

26. Lopez, D.S.; Wang, R.; Tsilidis, K.K.; Zhu, H.; Daniel, C.R.; Sinha, A.; Canfield, S. Role of Caffeine Intake on Erectile Dysfunction in US Men: Results from NHANES 2001–2004. *PLoS ONE* **2015**, *28*, e0123547. [CrossRef]

27. Chew, S.K.; Taouk, Y.; Xie, J.; Nicolaou, T.E.; Wang, J.J.; Wong, T.Y.; Lamoureux, E.L. Relationship between diabetic retinopathy, diabetic macular oedema and erectile dysfunction in type 2 diabetics. *Clin. Exp. Ophthalmol.* **2013**, *41*, 683–689. [CrossRef]

28. Berry, R.B.; Budhiraja, R.; Gottlieb, D.J.; Gozal, D.; Iber, C.; Kapur, V.K.; Marcus, C.L.; Mehra, R.; Parthasarathy, S.; Quan, S.F.; et al. American Academy of Sleep Medicine. Rules for scoring respiratory events in sleep: Update of the 2007 AASM Manual for the Scoring of Sleep and Associated Events. Deliberations of the Sleep Apnea Definitions Task Force of the American Academy of Sleep Medicine. *J. Clin. Sleep Med.* **2012**, *8*, 597–619.

29. Broström, A.; Johansson, P.; Albers, J.; Wiberg, J.; Svanborg, E.; Fridlund, B. 6-month CPAP-treatment in a young male patient with severe obstructive sleep apnoea syndrome—A case study from the couple's perspective. *Eur. J. Cardiovasc. Nurs.* **2008**, *7*, 103–112. [CrossRef]

30. Broström, A.; Johansson, P.; Strömberg, A.; Albers, J.; Mårtensson, J.; Svanborg, E. Obstructive sleep apnoea syndrome-patients' perceptions of their sleep and its effects on their life situation. *J. Adv. Nurs.* **2007**, *57*, 318–327. [CrossRef]

31. Stålkrantz, A.; Broström, A.; Wiberg, J.; Svanborg, E.; Malm, D. Everyday life for the spouses of patients with untreated OSA syndrome. *Scand. J. Caring Sci.* **2012**, *26*, 324–332. [CrossRef]

32. Hanak, V.; Jacobson, D.; McGree, M.; Sauver, J.S.; Lieber, M.M.; Olson, E.J.; Somers, V.K.; Gades, N.M.; Jacobsen, S.J. Snoring as a risk factor for sexual dysfunction in community men. *J. Sex Med.* **2008**, *5*, 898–908. [CrossRef]

33. Seftel, A.D.; Strohl, K.P.; Loye, T.L.; Bayard, D.; Kress, J.; Netzer, N.C. Erectile dysfunction and symptoms of sleep disorders. *Sleep* **2002**, *25*, 643–647. [CrossRef] [PubMed]

34. Husnu, T.; Ersoz, A.; Bulent, E.; Tacettin, O.; Remzi, A.; Bulent, A.; Aydin, M. Obstructive sleep apnea syndrome and erectile dysfunction: Does long term continuous positive airway pressure therapy improve erections? *Afr. Health Sci.* **2015**, *15*, 171–179. [CrossRef] [PubMed]

35. Szymański, F.; Puchalski, B.; Filipiak, K. Obstructive sleep apnea, atrial fibrillation and erectile dysfunction: Are they only coexisting conditions or a new clinical syndrome? The concept of the OSAFED syndrome. *Pol. Arch. Med. Wewn.* **2013**, *123*, 701–707. [CrossRef] [PubMed]

36. Andersen, M.L.; Santos-Silva, R.; Bittencourt, L.R.; Tufik, S. Prevalence of erectile dysfunction complaints associated with sleep disturbances in Sao Paulo, Brazil: A population-based survey. *Sleep Med.* **2010**, *11*, 1019–1024. [CrossRef]

37. Teloken, P.E.; Smith, E.B.; Lodowsky, C.; Freedom, T.; Mulhall, J.P. Defining association between sleep apnea syndrome and erectile dysfunction. *Urology* **2006**, *67*, 1033–1037. [CrossRef]

38. Luboshitzky, R.; Aviv, A.; Hefetz, A.; Herer, P.; Shen-Orr, Z.; Lavie, L.; Lavie, P. Decreased pituitary-gonadal secretion in men with obstructive sleep apnea. *J. Clin. Endocrinol. Metab.* **2002**, *87*, 3394–3398. [CrossRef]

39. Somers, V.K.; Dyken, M.E.; Clary, M.P.; Abboud, F.M. Sympathetic neural mechanisms in obstructive sleep apnea. *J. Clin. Investig.* **1995**, *96*, 1897–1904. [CrossRef]
40. Wolf, J.; Lewicka, J.; Narkiewicz, K. Obstructive sleep apnea: An update on mechanisms and cardiovascular consequences. *Nutr. Metab. Cardiovasc. Dis.* **2007**, *17*, 233–240. [CrossRef]
41. Christou, K.; Markoulis, N.; Moulas, A.N.; Pastaka, C.; Gourgoulianis, K.I. Reactive oxygen metabolites (ROMs) as an index of oxidative stress in obstructive sleep apnea patients. *Sleep Breath* **2003**, *7*, 105–110. [CrossRef] [PubMed]
42. Gonçalves, M.A.; Guilleminault, C.; Ramos, E.; Palha, A.; Paiva, T. Erectile dysfunction, obstructive sleep apnea syndrome and nasal CPAP treatment. *Sleep Med.* **2005**, *6*, 333–339. [CrossRef]
43. Jeon, Y.J.; Yoon, D.W.; Han, D.H.; Won, T.B.; Kim, D.Y.; Shin, H.W. Low Quality of Life and Depressive Symptoms as an Independent Risk Factor for Erectile Dysfunction in Patients with Obstructive Sleep Apnea. *J. Sex Med.* **2015**, *12*, 2168–2177. [CrossRef] [PubMed]
44. Madaeva, I.M.; Berdina, O.N.; Semenova, N.V.; Madaev, V.V.; Rychkova, L.V.; Kolesnikova, L.I. Sindrom obstruktivnogo apnoé sna i vozrastnoĭ gipogonadizm [Obstructive sleep apnea syndrome and age-related hypohonadism]. *Zhurnal Nevrol. Psikhiatrii Im. SS Korsakova.* **2017**, *117*, 79–83. (In Russian) [CrossRef] [PubMed]
45. Feng, C.; Yang, Y.; Chen, L.; Guo, R.; Liu, H.; Li, C.; Wang, Y.; Dong, P.; Li, Y. Prevalence and Characteristics of Erectile Dysfunction in Obstructive Sleep Apnea Patients. *Front. Endocrinol.* **2022**, *13*, 812974. [CrossRef]
46. Konstantinovsky, A.; Tamir, S.; Katz, G.; Tzischinsky, O.; Kuchersky, N.; Blum, N.; Blum, A. Erectile Dysfunction, Sleep Disorders, and Endothelial Function. *Isr. Med. Assoc. J.* **2019**, *21*, 408–411.
47. Hirshkowitz, M.; Whiton, K.; Albert, S.M.; Alessi, C.; Bruni, O.; DonCarlos, L.; Hazen, N.; Herman, J.; Adams Hillard, P.J.; Katz, E.S.; et al. National Sleep Foundation's updated sleep duration recommendations: Final report. *Sleep Health* **2015**, *1*, 233–243. [CrossRef]
48. Itani, O.; Jike, M.; Watanabe, N.; Kaneita, Y. Short sleep duration and health outcomes: A systematic review, meta-analysis, and meta-regression. *Sleep Med.* **2017**, *32*, 246–256. [CrossRef]
49. Dashti, H.S.; Scheer, F.A.; Jacques, P.F.; Lamon-Fava, S.; Ordovás, J.M. Short sleep duration and dietary intake: Epidemiologic evidence, mechanisms, and health implications. *Adv. Nutr.* **2015**, *6*, 648–659. [CrossRef] [PubMed]
50. National Center for Health Statistics. Quick-Stats: Percentage of adults who reported an average of $\leq$6 h of sleep per 24-h period, by sex and age group—United States, 1985 and 2004. *MMWR Morb. Mortal. Wkly. Rep.* **2005**, *54*, 933.
51. Kronholm, E.; Partonen, T.; Laatikainen, T.; Peltonen, M.; Härmä, M.; Hublin, C.; Kaprio, J.; Aro, A.R.; Partinen, M.; Fogelholm, M.; et al. Trends in self-reported sleep duration and insomnia-related symptoms in Finland from 1972 to 2005, a comparative review and re-analysis of Finnish population samples. *J. Sleep Res.* **2008**, *17*, 54–62. [CrossRef] [PubMed]
52. Jean-Louis, G.; Williams, N.J.; Sarpong, D.; Pandey, A.; Youngstedt, S.; Zizi, F.; Ogedegbe, G. Associations between inadequate sleep and obesity in the US population: Analysis of national health interview survey (1977–2009). *BMC Public Health* **2014**, *14*, 290. [CrossRef] [PubMed]
53. Vgontzas, A.N.; Fernandez-Mendoza, J.; Liao, D.; Bixler, E.O. Insomnia with objective short sleep duration: The most biologically severe phenotype of the disorder. *Sleep Med. Rev.* **2013**, *17*, 241–254. [CrossRef]
54. Zhang, F.; Xiong, Y.; Qin, F.; Yuan, J. Short Sleep Duration and Erectile Dysfunction: A Review of the Literature. *Nat. Sci. Sleep* **2022**, *27*, 1945–1961. [CrossRef] [PubMed]
55. Zou, Z.; Lin, H.; Zhang, Y.; Wang, R. The Role of Nocturnal Penile Tumescence and Rigidity (NPTR) Monitoring in the Diagnosis of Psychogenic Erectile Dysfunction: A Review. *Sex Med. Rev.* **2019**, *7*, 442–454. [CrossRef] [PubMed]
56. Ai, S.; Zhang, J.; Zhao, G.; Wang, N.; Li, G.; So, H.-C.; Liu, Y.; Chau, S.W.-H.; Chen, J.; Tan, X.; et al. Causal associations of short and long sleep durations with 12 cardiovascular diseases: Linear and nonlinear Mendelian randomization analyses in UK Biobank. *Eur. Heart J.* **2021**, *42*, 3349–3357. [CrossRef]
57. Grandner, M.A.; Patel, N.P.; Gehrman, P.R.; Perlis, M.L.; Pack, A.I. Problems associated with short sleep: Bridging the gap between laboratory and epidemiological studies. *Sleep Med. Rev.* **2010**, *14*, 239–247. [CrossRef]

Systematic Review

Artificial Intelligence as an Aid in CBCT Airway Analysis: A Systematic Review

Ioannis A. Tsolakis [1,*], Olga-Elpis Kolokitha [1], Erofili Papadopoulou [2], Apostolos I. Tsolakis [3,4], Evangelos G. Kilipiris [5] and J. Martin Palomo [4]

1 Department of Orthodontics, School of Dentistry, Aristotle University of Thessaloniki, 54124 Thessaloniki, Greece
2 Department of Oral Medicine & Pathology and Hospital Dentistry, School of Dentistry, National and Kapodistrian University of Athens, 10679 Athens, Greece
3 Department of Orthodontics, School of Dentistry, National and Kapodistrian University of Athens, 10679 Athens, Greece
4 Department of Orthodontics, Case Western Reserve University School of Dental Medicine, Cleveland, OH 44106, USA
5 Department of Maxillofacial Surgery, F.D. Roosevelt University Hospital, 97517 Banska Bystrica, Slovakia
* Correspondence: tsolakisioannis@gmail.com

Abstract: Background: The use of artificial intelligence (AI) in health sciences is becoming increasingly popular among doctors nowadays. This study evaluated the literature regarding the use of AI for CBCT airway analysis. To our knowledge, this is the first systematic review that examines the performance of artificial intelligence in CBCT airway analysis. Methods: Electronic databases and the reference lists of the relevant research papers were searched for published and unpublished literature. Study selection, data extraction, and risk of bias evaluation were all carried out independently and twice. Finally, five articles were chosen. Results: The results suggested a high correlation between the automatic and manual airway measurements indicating that the airway measurements may be automatically and accurately calculated from CBCT images. Conclusions: According to the present literature, automatic airway segmentation can be used for clinical purposes. The main key findings of this systematic review are that the automatic airway segmentation is accurate in the measurement of the airway and, at the same time, appears to be fast and easy to use. However, the present literature is really limited, and more studies in the future providing high-quality evidence are needed.

Keywords: Artificial Intelligence; CBCT; airway

Citation: Tsolakis, I.A.; Kolokitha, O.-E.; Papadopoulou, E.; Tsolakis, A.I.; Kilipiris, E.G.; Palomo, J.M. Artificial Intelligence as an Aid in CBCT Airway Analysis: A Systematic Review. *Life* **2022**, *12*, 1894. https://doi.org/10.3390/life12111894

Academic Editor: Konstantinos Chaidas

Received: 11 October 2022
Accepted: 11 November 2022
Published: 15 November 2022

Publisher's Note: MDPI stays neutral with regard to jurisdictional claims in published maps and institutional affiliations.

1. Introduction

The digital era in health sciences has been ushered in by recent innovations like cone beam computed tomography (CBCT), 3D printing, and artificial intelligence (AI). Those innovations have been playing an important role in the field of health sciences to support diagnosis and customized treatment solutions [1–3]. In 1956, Dartmouth University was the first to use the terminology "artificial intelligence", which is defined as computerized synthetic human cognitive function [4]. Since then, the application of AI has grown dramatically [5,6]. Thanks to improvements in analytics methods, computing power, and data accessibility, AI may touch many aspects of modern culture. On a global scale, we are already noticing its effects on our day-to-day lives. In addition to filtering information for social media and web searches, it also does this for consumer electronics like cameras, cellphones, tablets, and even cars. Due to the scientific method, which is presently undergoing a paradigm shift in the multidisciplinary link between AI and data science, all recent advancements and advances in dentistry have been made possible [7,8].

Artificial intelligence (AI) refers to fundamental technologies including deep learning, artificial neural networks (ANNs), and machine learning. A significant area of artificial intelligence is machine learning. Machine learning makes predictions about new data and circumstances using the statistical patterns of previously learned data. Training data are necessary for machine learning to function. Machine learning cannot work without training data. With this approach, the computer model can develop over time by learning from experience rather than through conventional explicit programming. A model needs a large quantity of data to using abstractions from various processing levels. A subset of artificial intelligence is deep learning. Deep learning is the process by which computers learn to think utilizing structures inspired by the human brain, whereas machine learning is the process by which computers learn to think and act with less human intervention. The benefit of deep learning is that little engineering effort is needed to prepare the data for analysis. Recognition of visual objects and item identification have seen the most use of deep learning techniques. Orthodontic and Otolaryngology (ORL) clinical applications favor more advanced AI solutions, such as cone beam computed tomography (CBCT) and 3D convolutional neural networks (3D CNN). "Strong AI" or "deep AI" is a type of artificial general intelligence (AGI) that is comparable to humans in terms of problem-solving ability. In contrast to the majority of today's advanced AI algorithms, physicians possess the capacities for abstract thought, strategic planning, and the generation of original ideas [9]. AGI will have the capacity to think very much like a human. Beyond our comprehension, artificial super intelligence (ASI) will be able to learn and grow beyond human capacities.

A notable dilemma involving legal culpability for flaws in AI algorithm assessment and potential erroneous medical intervention is one of the issues surrounding AI application. As is well known, AI can be programmed to reflect the biases of any individual [10]. We do not know how deep AI algorithms generated the results, which is another dilemma that Zhang et al. addressed [11]. Due to the black-box nature of AI processes, current research has focused on "explainable artificial intelligence (XAI)" to get around this limitation. In contrast to AI methods like deep learning, XAI may offer both decision-making and model explanations [11]. Several publications cover this subject [12,13].

The upper airway, also known as the pharyngeal airway space, is a complicated anatomical region that is closely related to the nearby bone and soft tissue components. It is mostly in charge of carrying out actions including breathing, speaking, and swallowing [14]. Since multiple investigations have shown its connection to craniofacial growth and development, the upper airway assessment has attracted the attention of physicians [15–18]. Since the craniofacial complex could be responsible for possible constrictions of the upper airway, physicians used surgical and non-invasive methods to change the anatomy and resolve the airway constriction. The use of X-rays is really important to assess the effectiveness and find the possible side effects of these treatments [19]. In the past, two-dimensional (2D) lateral cephalometry was used to evaluate airway alterations in patients with dentofacial and skeletal anomalies during the stages of diagnosis, treatment planning, and follow-up [20,21]. However, due to the possible drawbacks of 2D approaches to representing the upper airway, computed tomography (CT), cone-beam CT (CBCT), and magnetic resonance imaging (MRI) have largely taken their position as a clinical standard for assessing upper airway volume and dimensional changes in order to comprehend their pathogenesis [22–24]. CBCT has proven to be as accurate as other gold standard methods for measuring the upper airway volume and constricted area [25].

It is now proved that CBCT can be used in every day practice to accurately measure the airway volume and minimum cross-section area; this plays a critical role in evaluating and managing different airway disorders [25]. The use of artificial intelligence in CBCT airway analysis could provide accurate and fast measurements to the clinicians. This would be translated to faster management of different airway disorders that could be life threatening. This article aims to systematically review the current literature on the use of artificial intelligence in CBCT airway analysis. To our knowledge, this is the first systematic review that examines the performance of artificial intelligence in CBCT airway analysis.

2. Materials and Methods

2.1. Protocol and Registration

The protocol for this present systematic review was registered on the Open Science Forum Database following the Prisma-P guidelines1 (Protocol: 10.17605/OSF.IO/4HWBJ).

This systematic review was conducted by using the following keywords in the search strategy "Artificial Intelligence", "airway volume", "cbct". Those keywords were combined with the following Medical Subject Heading (MeSH terms): "Artificial Intelligence" [Mesh], "Cone-Beam Computed Tomography" [Mesh]. The databases used for the electronic search were Med-line (PubMed), Cochrane Library, and Scopus. A manual search was also carried out. There was a choice of exclusively English-written papers, and the publication duration was unrestricted. Personal opinions were omitted from studies. The search was conducted for studies published until July 2022. The search strategy for PubMed is presented in Table 1.

Table 1. The search strategy for PubMed.

"Cone-Beam Computed Tomography" [Mesh] AND airway volume	330 results
"Cone-Beam Computed Tomography" [Mesh] AND Artificial Intelligence" [Mesh]	257 results
Artificial Intelligence" [Mesh] AND airway volume	76 results
"Cone-Beam Computed Tomography" [Mesh] AND Artificial Intelligence" [Mesh] AND airway volume	4 results

Studies were chosen by three authors separately and in duplication (I.A.T., E.P., E.G.K.). Discussion with other authors helped to clarify any potential inconsistencies (O.K., A.I.T, J.M.P). The names of the studies' authors, their institutions, or their research conclusions were revealed (not blinded). The authors first looked for possibly pertinent research by title, then they read the abstract and eliminated any irrelevant papers. Later, to locate more papers that were missed by database searches, a manual search of relevant study references was conducted. Finally, after thoroughly reviewing all of the papers, a decision was taken based on our inclusion and exclusion criteria (Table 2).

Table 2. Inclusion and Exclusion criteria.

Inclusion Criteria	Exclusion Criteria
Studies that refer to the use of artificial intelligence for CBCT airway analysis prospective or retrospective studies	Studies that are reviews or authors' opinion

2.2. Data Items and Collection Extraction and Management

The data were independently extracted and duplicated by three review writers (I.A.T., E.P., E.G.K.). Study participants, the intervention, the results, the techniques of outcome evaluation, the findings, and the conclusion were among the data that were extracted. The present authors reported and examined only the data that were available because they did not have access to any missing data.

2.3. Risk of Bias/Quality Assessment in Individual Studies

The Cochrane Quality Assessment of the ACROBAT-NRSI tool was used to evaluate the methodology of the included studies and determine whether there were any applicability or bias problems. Based on the following, each domain was evaluated and classified as high risk, low risk, or unclear:

1. Low risk of bias if all key domains of the study were at low risk of bias.
2. Unclear risk of bias if one or more key domains of the study were unclear.
3. High risk of bias if one or more key domains were at high risk of bias.

3. Results

In the initial data search, 1050 studies were found from all data bases. Only 70 of these papers were chosen, based on the study's title. Each chosen article was then thoroughly assessed by three independent authors who read the complete document. Five publications in all were chosen for the present systematic review. Four studies were the result of PubMed search while on extra research paper was found through Scopus.

All the final selected articles were prospective studies. All studies evaluated the accuracy of AI systems in segmenting and calculating airway volume based on CNN and RNN models. Three articles used their own model for software usage while one of the remaining two used the Mimics 19.0, InVivo 5 software and the other one the Diognocat, InVivo 5 software [26–30]. The procedure of article selection is presented on a flow diagram (Figure 1), and data are briefly presented in Table 3.

Risk of Bias within Studies

The following seven criteria were applied to non-RCT studies: bias due to confounding, bias in the selection of participants into the study, bias in the measurement of interventions, bias due to departures from intended interventions, bias due to missing data, bias in measurement outcomes and bias in the selection of the reported result. Two of the studies presented high risk of bias while the other three presented low risk of bias in all measurements (Table 4).

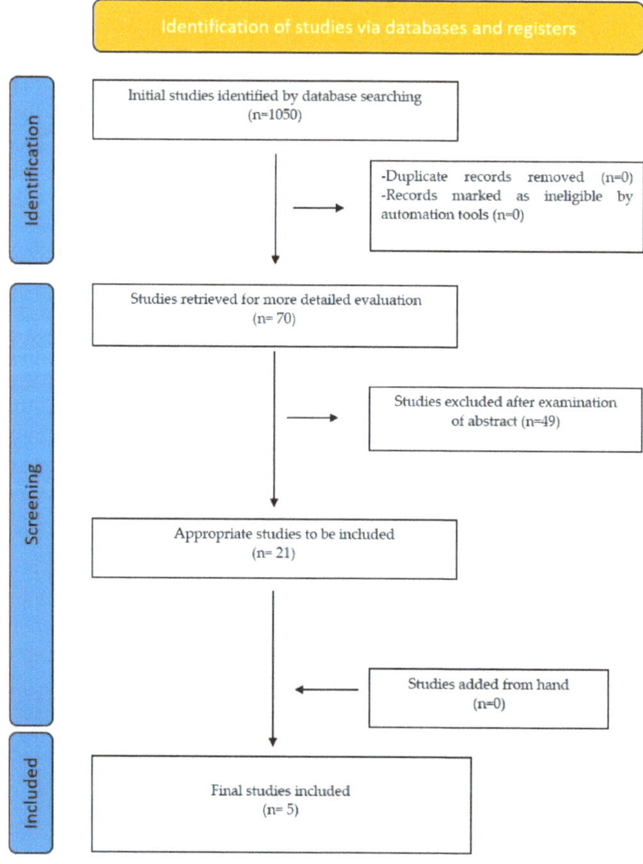

Figure 1. Prisma Flow diagram–selection of studies.

Table 3. Data extraction.

Authors/ Publication Year	Study Design	Participants (Number of CBCT)	Intervention	Outcomes	Method of Outcome Assessment		Results	Authors/ Publication Year
Leonardi R et al. [26] (2021)	prospective	40 CBCT scans	1 CBCT device was used.	Accuracy of the CNN fully automatic segmentation of the sinonasal cavity pharyngeal airway	-	Mimics software (version 22.0, materialize N.V., Leuven, Belgium). Surface-to-surface matching technique.	• Measurements were highly correlated with an intraclass correlation coefficient value of 0.921, whereas the method error was 0.31 mm^3. • A mean difference of $1.93 \pm 6 \pm 0.73$ mm^3 was found to be not statistically significant. • The differences, measured as the Dice score coefficient in percentage, between the assessments done with both methods were 3.3% and 5.8%, respectively.	The new deep learning–based method for automated segmentation is accurate for airway segmentation
Park et al. [27]	prospective	315 CBCT scans	1 CBCT device was used.	Accuracy of the airway volume measurement by a Regression Neural Network-based deep-learning model	-	MATLAB 2020a (MathWorks, Natick, MA, USA)	• The total volume was the most correlated intra-class correlation coefficient (ICC) value in the oropharynx (0.986), followed by the hypopharynx (0.964), and the nasopharynx (0.912). • The slope of the two measurements was close to 1 and showed a linear regression correlation ($r^2 = 0.975$, slope = 1.02, p < 0.001).	These results indicate that fully automatic segmentation of the airway is possible by training via deep learning of artificial intelligence.
Shujaat S et al. [28] (2021)	prospective	103 CT and CBCT scans	Scans from 1 CT and 2 CBCTs were grouped in: - training set - test set - validation set	The performance of deep learning based 3D CNN model for automatic segmentation of the pharyngeal airway space	-	Mimics software (version 22.0, materialize N.V., Leuven, Belgium).	• The CNN model was able to identify the segmented region with optimal precision and recall. • The maximal difference between the automatic segmentation and ground truth was 0.98 ± 0.74 mm. • The dice score of 0.97 ± 0.02 confirmed the high similarity of the segmented region to the ground truth.	The proposed 3D U-Net model offered an accurate method for the segmentation of Airway from CT/CBCT images.
Sin Ç et al. [29] (2021)	prospective	306 CBCT scans	1 CBCT device was used and grouped in: - training, - validation - test sets	The accuracy of an automatic detection algorithm for pharyngeal airway on CBCT images using a deep-learning artificial intelligence system	-	Open-source version 3.8 ITK SNAP software - MATLAB implementation of U-Net and SGD Adam optimizer.	• There was no statistically significant difference between the human observation of the average volume of the pharyngeal airway and the results from artificial intelligence • The ICC between researcher and AI measurements was found to be highly correlated (0.985) • The calculated Dice ratio across all slices of all CBCT images was 0.919, and the mean accuracy of 0.961 providing excellent accuracy.	AI models based on deep learning techniques can be used for easy and error-free segmentation of pharyngeal airway volume from CBCT
Orhan K et al. [30] (2022)	prospective	200 CBCT scans	3 CBCT devices	To validate an automatic detection algorithm for pharyngeal airway on CBCT data using an AI software for OSA patients. To validate the newly developed artificial intelligence system in comparison to commercially available software for 3D CBCT evaluation.	-	Diagnocat, InVivo 5	• There was no statistically significant difference between the manual technique and Diagnocat measurements in all groups (p > 0.05). • Inter-class correlation coefficients were 0.954 for manual and automatic segmentation, 0.956 for Diagnocat and automatic segmentation, 0.972 for Diagnocat and manual segmentation. • It was seen that the DC algorithm also measures the epiglottis volume and the posterior nasal aperture volume due to the low soft-tissue contrast in CBCT images; this leads to higher values in airway volume measurement.	Activating this potential collaboration for OSA patients would significantly reduce the effort and time required for the initial diagnosis and follow-up of these patients.

Table 4. Risk of bias assessment.

Author (Year)	Outcomes	Bias Due to Confounding	Bias in Selection of Participants into the Study	Bias in Measurement of Interventions	Bias Due to Departures from Intended Interventions	Bias Due to Missing Data	Bias in Measurement of Outcomes	Bias in Selection of the Reported Result	Overall Bias
Leonardi R et al. [26] (2021)	• Accuracy of the CNN fully automatic segmentation of the airway	Low for all outcomes	Low for all outcomes	Low for all outcomes	Low for all outcomes	Low for all outcomes	High for all outcomes	Low for all outcomes	High for all outcomes
Park et al. [27] (2021)	• Accuracy of the airway volume measurement by a Regression Neural Network	Low for all outcomes	Low for all outcomes	Low for all outcomes	Low for all outcomes	Low for all outcomes	High for all outcomes	Low for all outcomes	High for all outcomes
Shujaat S et al. [28] (2021)	• Accuracy of 3D CNN model for automatic segmentation of the pharyngeal airway space	Low for all outcomes	Low for all outcomes	Low for all outcomes	Low for all outcomes	Low for all outcomes	Low for all outcomes	Low for all outcomes	Low for all outcomes
Sin Ç et al. [29] (2021)	• Accuracy of an automatic detection algorithm for pharyngeal airway	Low for all outcomes	Low for all outcomes	Low for all outcomes	Low for all outcomes	Low for all outcomes	Low for all outcomes	Low for all outcomes	Low for all outcomes
Orhan K et al. [30] (2022)	• To validate an automatic detection algorithm for pharyngeal airway with AI for OSA patients	Low for all outcomes	Low for all outcomes	Low for all outcomes	Low for all outcomes	Low for all outcomes	Low for all outcomes	Low for all outcomes	Low for all outcomes

4. Discussion

All studies in the present systematic review reported equal conclusions regarding the accuracy and the benefit of automatic segmentation of the upper airway by deep learning methods. It is important to mention that all studies were reported in 2021 and after. This shows that the scientific interest in AI and airway volume segmentation has really increased in the last few years. We were able to access all the known databases to minimize the limitations of this study. However, a possible limitation could appear for articles not included in any of the databases we used.

The first and most important stage in assessing the upper airway volume is segmentation, which allows the airway space to be distinguished from the rest of the scan and can then be visualized and quantified in three dimensions. Various upper airway segmentation methods and algorithms that are either manual, semiautomatic, or automatic in nature have been developed over the past ten years [31]. Even while hand segmentation is the gold standard and provides the most precise replication of the anatomical structure, it is labor- and time-intensive. Numerous threshold-based semiautomatic software programs have been approved for volumetric assessment as well [32]. In these programs, the user specifies a volume of interest (VOI), and the program automatically combines the gray threshold values in that area without effectively taking into account the image intensity and anatomical variations. Similar to this, several studies have proposed a set threshold value for segmenting the upper airway [33,34], although this value may change based on the CBCT equipment, scanning parameters, machine calibration, and noise from patient movement or metal artifacts [35,36]. The semiautomatic method has proven to be accurately measured by different software [37]. For the segmentation of the upper airway, prior studies have also presented fully automatic advanced and hybrid picture segmentation techniques. However, due to either poor precision, fixed thresholding, manual localization of seed points, manual VOI selection, reliance on picture orientation, or algorithmic failure under varied scanning parameters, they are of limited value [32,38–40].

The first study discussed here was reported in early 2021 from Leonardi et al. Aiming to fully automate the segmentation of the pharyngeal airway, and the sinonasal cavity from cone-beam computed tomography (CBCT) scans, they evaluated the accuracy of a new autonomous deep learning-based approach. In order to manually segment the sinonasal cavity and pharyngeal subregions of 40 healthy patients' CBCT scans (20 women and 20 men), Mimics software was used (version 20.0; Materialise, Leuven, Belgium). A total of 20 CBCT scans were chosen at random from the entire sample and used to train the AI model file. By contrasting the segmentation volumes of the 3D models created with automatic and manual segmentations, the remaining 20 CBCT segmentation masks were utilized to assess the precision of the CNN completely automatic technique. Their results suggested a low model error (0.31 mm^3), and all the measurements were highly correlated with an intraclass correlation coefficient of 0.921. Between the approaches, there was a mean difference of 1.93 ± 0.73 cm^3, but it was not statistically significant ($p > 0.05$). The average matching percentage found was 85.35 ± 2.59 and 93.44 ± 2.54. The disparities between the assessments made using the two approaches were 3.3% and 5.8%, respectively, as expressed by the Dice score coefficient in percentage [26].

In April of 2021, Park et al. reported on the precision of an airway volume measurement made using a deep learning model based on regression neural networks. The system for entirely automatic segmentation of a deep learning process was built using a set of manually drawn airway data. One examiner used the mid-sagittal plane of 315 patients' cone-beam computed tomography (CBCT) scans to identify the manual landmarks of the airway. They performed clinical dataset-based training with data augmentation. The airway channel was measured and segmented using markers that were annotated. The authors were able to verify the accuracy of the model by assessing the following differences between the examiner and the program: (1) a difference in the nasopharynx, oropharynx, and hypopharynx volume; and (2) the Euclidean distance. A total of 61 samples were collected and compared for the agreement analysis. The correlation test revealed a reliability range from high to outstanding. Regression analysis was used to examine differences in volume. There was a strong linear regression connection between the slopes of the two measurements, which was close to 1 (r^2 = 0.975, slope = 1.02, p < 0.001). These findings suggested that fully automatic airway segmentation is trainable using deep learning in artificial intelligence. Furthermore, they found that there was a strong correlation between manual data and deep learning data [27].

A month later, and more specifically in May of 2021, Shujaat et al. reported an evaluation of a deep learning-based three-dimensional (3D) convolutional neural network (CNN) model for automatically segmenting the pharyngeal airway space, and its performance was examined. From a database of individuals undergoing orthognathic surgery, 103 CT and CBCT scans were obtained. Two CBCT devices (Promax 3D Max, Planmeca, Helsinki, Finland, and Newtom VGi evo, Cefla, Imola, Italy) and one CT (128-slice multi-slice spiral CT, Siemens Somatom Definition Flash, Siemens AG, Erlangen, Germany) with various scanning parameters made up the acquisition devices. The airway was automatically segmented using a 3D CNN-based model called the 3D U-Net. The entire CT/CBCT dataset was divided into three sets: training set (n = 48) for training the model based on the observer-based manual segmentation that provided the basis for it; test set (n = 25) for obtaining the model's final performance; and validation set (n = 30) for comparing the model's performance to that of observer-based segmentation. Their results suggested that the segmented region could be distinguished by the CNN model with the best precision (0.97 ± 0.01) and recall (0.96 ± 0.03). The maximum deviation between the ground truth and the artificial segmentation based on the 95% Hausdorff distance score was 0.98 ± 0.74 mm. The segmented region's high likeness to the real world was validated by the dice score of 0.97 ± 0.02. It was also discovered that the Intersection over Union (IoU) metric had a high value (0.93 ± 0.03). In comparison to the Promax 3D Max and CT device, the Newtom VGi Evo CBCT performed better based on the acquisition devices [28].

In December of 2021, Sin et al. reported on a deep learning artificial intelligence (AI) system to assess an automatic segmentation algorithm for the pharyngeal airway in cone-beam computed tomography (CBCT) images. In this retrospective investigation, data from 306 participants on the pharyngeal airway were included after archives of CBCT pictures were reviewed. Using serial CBCT images, a machine learning method built on Convolutional Neural Networks (CNN) segmented the pharyngeal airway. The airway was manually generated using semi-automatic software (ITK-SNAP), and the outcomes were contrasted with those of artificial intelligence. When comparing the results of measurements made by humans versus algorithms powered by artificial intelligence, the dice similarity coefficient (DSC) and intersection over union (IoU) were utilized as measures of segmentation accuracy. The average pharyngeal airway volume, according to the human observer, was 18.08 cm^3, while the average volume of artificial intelligence was 17.32 cm^3. It is possible to segment the pharyngeal airways with a dice ratio of 0.919 and a weighted IoU of 0.993 [29].

Finally, the study of Orhan et al. (2022) was characterized by two goals. The first goal of this work was to develop and verify an algorithm for automatically detecting the pharyngeal airway on CBCT data using artificial intelligence (AI) software called Diacat. The second goal was to compare the recently created artificial intelligence system to commercially available software for 3D CBCT evaluation in order to validate it. The pharyngeal airway in obstructive sleep apnea (OSA) patients was automatically assessed for the first time in this study. The segmentation of the pharyngeal airways in OSA and non-OSA patients was performed using a machine learning technique based on convolutional neural networks. Radiologists manually determined the airway using semi-automatic software, and their measurements were compared with those of the AI. The mean airway volumes of the several OSA patient groups (minimal, mild, moderate, and severe) were compared. In addition, patients with OSA and those without it were compared in terms of their airway's narrowest points (mm), its field (mm^2), and its volume (cc). In all groups, there was no statistically significant difference between measures taken using the manual method and those taken using the Diagnocat ($p > 0.05$). For manual and automated segmentation, the interclass correlation coefficients were 0.954 for Diagnocat, and 0.956 for automatic segmentation. They examined the output images to determine why the mean value for the total airway was higher in the DC measurement, even though there was no statistically significant difference in total airway volume measurements between the manual measurements, automatic measurements, and DC measurements in non-OSA and OSA patients. Due to the low soft-tissue contrast in CBCT images, it was shown that the DC algorithm also assesses the epiglottis volume and the posterior nasal aperture volume, which results in greater values for airway volume measurement [30].

According to this systematic review the present literature is really limited as regards studies that looked at the accuracy of artificial intelligence use for CBCT upper airway analysis. In order to have strong evidence on how accurately CBCT airway analysis is performed with the use of artificial intelligence, more randomized prospective studies should be performed. Those studies should limit all kind of bias in order to provide high quality evidence.

5. Conclusions

The pharyngeal airway may now be automatically segmented from CBCT images thanks to a successful AI algorithm. According to the present literature, the automatic segmentation can be put to clinical use. This is because it appears to be accurate in the measurement of the airway but at the same time it appears to be fast and easy to use. However, there were only 5 studies in the present literature to support these data and only 3 of them reported a low risk of bias. More studies in the future providing high-quality evidence are needed.

Author Contributions: I.A.T., data curation, formal analysis, investigation, methodology, resources, software, validation, visualization, roles/writing—original draft; O.-E.K., project administration, supervision, writing—review and editing; E.P., data curation, formal analysis, investigation, resources, writing—review and editing; A.I.T., project administration, supervision, writing—review and editing; E.G.K., data curation, formal analysis, investigation, resources, writing—review and editing; J.M.P., conceptualization, methodology, project administration, supervision, validation, writing—review and editing. All authors have read and agreed to the published version of the manuscript.

Funding: This research received no external funding.

Institutional Review Board Statement: Not applicable.

Informed Consent Statement: Not applicable.

Data Availability Statement: Not applicable.

Conflicts of Interest: The authors declare no conflict of interest.

References

1. Thurzo, A.; Urbanová, W.; Novák, B.; Czako, L.; Siebert, T.; Stano, P.; Mareková, S.; Fountoulaki, G.; Kosnáčová, H.; Varga, I. Where Is the Artificial Intelligence Applied in Dentistry? Systematic Review and Literature Analysis. *Healthcare* **2022**, *10*, 1269. [CrossRef] [PubMed]
2. Tsolakis, I.A.; Gizani, S.; Panayi, N.; Antonopoulos, G.; Tsolakis, A.I. Three-Dimensional Printing Technology in Orthodontics for Dental Models: A Systematic Review. *Children* **2022**, *9*, 1106. [CrossRef] [PubMed]
3. Tsolakis, I.A.; Gizani, S.; Tsolakis, A.I.; Panayi, N. Three-Dimensional-Printed Customized Orthodontic and Pedodontic Appliances: A Critical Review of a New Era for Treatment. *Children* **2022**, *9*, 1107. [CrossRef] [PubMed]
4. Obermeyer, Z.; Emanuel, E.J. Predicting the Future—Big Data, Machine Learning, and Clinical Medicine. *N. Engl. J. Med.* **2016**, *375*, 1216. [CrossRef] [PubMed]
5. Gharavi, S.M.H.; Faghihimehr, A. Clinical Application of Artificial Intelligence in PET Imaging of Head and Neck Cancer. *PET Clin.* **2022**, *17*, 65–76. [CrossRef]
6. Patil, S.; Albogami, S.; Hosmani, J.; Mujoo, S.; Kamil, M.A.; Mansour, M.A.; Abdul, H.N.; Bhandi, S.; Ahmed, S.S.S.J. Artificial Intelligence in the Diagnosis of Oral Diseases: Applications and Pitfalls. *Diagnostics* **2022**, *12*, 1029. [CrossRef]
7. Rasteau, S.; Ernenwein, D.; Savoldelli, C.; Bouletreau, P. Artificial Intelligence for Oral and Maxillo-Facial Surgery: A Narrative Review. *J. Stomatol. Oral Maxillofac. Surg.* **2022**, *123*, 276–282. [CrossRef]

8.   Monterubbianesi, R.; Tosco, V.; Vitiello, F.; Orilisi, G.; Fraccastoro, F.; Putignano, A.; Orsini, G. Augmented, Virtual and Mixed Reality in Dentistry: A Narrative Review on the Existing Platforms and Future Challenges. *Appl. Sci.* **2022**, *12*, 877. [CrossRef]
9.   Oshida, Y. Artificial Intelligence for Medicine. *Artif. Intell. Med.* **2021**, *69*, S36–S40.
10. Obermeyer, Z.; Powers, B.; Vogeli, C.; Mullainathan, S. Dissecting Racial Bias in an Algorithm Used to Manage the Health of Populations. *Science* **2019**, *366*, 447–453. [CrossRef]
11. Zhang, Y.; Weng, Y.; Lund, J.; Faust, O.; Su, L.; Acharya, R. Applications of Explainable Artificial Intelligence in Diagnosis andSurgery. *Diagnostics* **2022**, *12*, 237. [CrossRef] [PubMed]
12. Kavya, R.; Christopher, J.; Panda, S.; Lazarus, Y.B. Machine Learning and XAI Approaches for Allergy Diagnosis. *Biomed. Signal Process. Control* **2021**, *69*, 102681. [CrossRef]
13. Barredo Arrieta, A.; Díaz-Rodríguez, N.; del Ser, J.; Bennetot, A.; Tabik, S.; Barbado, A.; Garcia, S.; Gil-Lopez, S.; Molina, D.; Benjamins, R.; et al. Explainable Artificial Intelligence (XAI): Concepts, Taxonomies, Opportunities and Challenges toward Responsible AI. *Inf. Fusion* **2020**, *58*, 82–115. [CrossRef]
14. Di Carlo, G.; Polimeni, A.; Melsen, B.; Cattaneo, P.M. The relationship between upper airways and craniofacial morphology studied in 3D. A CBCT study. *Orthod. Craniofac. Res.* **2015**, *18*, 1–11. [CrossRef]
15. Schendel, S.A.; Jacobson, R.; Khalessi, S. Airway growth and development: A computerized 3-dimensional analysis. *J. Oral Maxillofac. Surg.* **2012**, *70*, 2174–2183. [CrossRef]
16. Claudino, L.V.; Mattos, C.T.; Ruellas, A.C.D.O.; Sant Anna, E.F. Pharyngeal airway characterization in adolescents related to facial skeletal pattern: A preliminary study. *Am. J. Orthod. Dentofac. Orthop.* **2013**, *143*, 799–809. [CrossRef]
17. Zheng, Z.H.; Yamaguchi, T.; Kurihara, A.; Li, H.F.; Maki, K. Three-dimensional evaluation of upper airway in patients with different anteroposterior skeletal patterns. *Orthod. Craniofac. Res.* **2014**, *17*, 38–48. [CrossRef]
18. Celikoglu, M.; Bayram, M.; Sekerci, A.E.; Buyuk, S.K.; Toy, E. Comparison of pharyngeal airway volume among different vertical skeletal patterns: A cone-beam computed tomography study. *Angle Orthod.* **2014**, *84*, 782–787. [CrossRef]
19. Tsolakis, I.A.; Palomo, J.M.; Matthaios, S.; Tsolakis, A.I. Dental and Skeletal Side Effects of Oral Appliances Used for the Treatment of Obstructive Sleep Apnea and Snoring in Adult Patients—A Systematic Review and Meta-Analysis. *J. Pers. Med.* **2022**, *12*, 483. [CrossRef]
20. Pereira-Filho, V.A.; Castro-Silva, L.M.; De Moraes, M.; Gabrielli, M.F.R.; Campos, J.A.D.B.; Juergens, P. Cephalometric evaluation of pharyngeal airway space changes in class III patients undergoing orthognathic surgery. *J. Oral Maxillofac. Surg.* **2011**, *69*, e409–e415. [CrossRef]
21. Gungor, A.Y.; Turkkahraman, H.; Yilmaz, H.H.; Yariktas, M. Cephalometric comparison of obstructive sleep apnea patients and healthy controls. *Eur. J. Dent.* **2013**, *7*, 48–54. [CrossRef] [PubMed]
22. Aboudara, C.; Nielsen, I.; Huang, J.C.; Maki, K.; Miller, A.J.; Hatcher, D. Comparison of airway space with conventional lateral headfilms and 3-dimensional reconstruction from cone-beam computed tomography. *Am. J. Orthod. Dentofac. Orthop.* **2009**, *135*, 468–479. [CrossRef] [PubMed]
23. He, J.; Wang, Y.; Hu, H.; Liao, Q.; Zhang, W.; Xiang, X.; Fan, X. Impact on the upper airway space of different types of orthognathic surgery for the correction of skeletal class III malocclusion: A systematic review and meta-analysis. *Int. J. Surg.* **2017**, *38*, 31–40. [CrossRef] [PubMed]
24. Alsufyani, N.A.; Noga, M.L.; Witmans, M.; Major, P.W. Upper airway imaging in sleep-disordered breathing: Role of cone-beam computed tomography. *Oral Radiol.* **2017**, *33*, 161–169. [CrossRef]
25. Tsolakis, I.A.; Venkat, D.; Hans, M.G.; Alonso, A.; Palomo, J.M. When static meets dynamic: Comparing cone-beam computed tomography and acoustic reflection for upper airway analysis. *Am. J. Orthod. Dentofac. Orthop.* **2016**, *150*, 643–650. [CrossRef]
26. Leonardi, R.; Lo Giudice, A.; Farronato, M.; Ronsivalle, V.; Allegrini, S.; Musumeci, G.; Spampinato, C. Fully automatic segmentation of sinonasal cavity and pharyngeal airway based on convolutional neural networks. *Am. J. Orthod. Dentofac. Orthop.* **2021**, *159*, 824–835.e1. [CrossRef]
27. Park, J.; Hwang, J.; Ryu, J.; Nam, I.; Kim, S.-A.; Cho, B.-H.; Shin, S.-H.; Lee, J.-Y. Deep Learning Based Airway Segmentation Using Key Point Prediction. *Appl. Sci.* **2021**, *11*, 3501. [CrossRef]
28. Shujaat, S.; Jazil, O.; Willems, H.; Van Gerven, A.; Shaheen, E.; Politis, C.; Jacobs, R. Automatic segmentation of the pharyngeal airway space with convolutional neural network. *J. Dent.* **2021**, *111*, 103705. [CrossRef]
29. Sin, Ç.; Akkaya, N.; Aksoy, S.; Orhan, K.; Öz, U. A deep learning algorithm proposal to automatic pharyngeal airway detection and segmentation on CBCT images. *Orthod. Craniofac. Res.* **2021**, *24*, 117–123. [CrossRef]
30. Orhan, K.; Shamshiev, M.; Ezhov, M.; Plaksin, A.; Kurbanova, A.; Ünsal, G.; Gusarev, M.; Golitsyna, M.; Aksoy, S.; Mısırlı, M.; et al. AI-based automatic segmentation of craniomaxillofacial anatomy from CBCT scans for automatic detection of pharyngeal airway evaluations in OSA patients. *Sci. Rep.* **2022**, *12*, 118. [CrossRef]
31. El Khateeb, S. Three-dimensional image segmentation of upper airway by cone beam CT: A review of literature. *Egypt. Dent. J.* **2020**, *66*, 1527–1535. [CrossRef]
32. De Menezes Weissheimer, L.M.E.; Sameshima, G.T.; Enciso, R.; Pham, J.; Grauer, D. Imaging software accuracy for 3-dimensional analysis of the upper airway. *Am. J. Orthod. Dentofac. Orthop.* **2012**, *142*, 801–813. [CrossRef]
33. Lenza, M.G.; Lenza, M.D.O.; Dalstra, M.; Melsen, B.; Cattaneo, P.M. An analysis of different approaches to the assessment of upper airway morphology: A CBCT study. *Orthod. Craniofacial Res.* **2010**, *13*, 96–105. [CrossRef] [PubMed]

34. Alves, M.; Baratieri, C.; Mattos, C.T.; Brunetto, D.; Fontes, R.D.C.; Santos, J.R.L.; Ruellas, A.C.D.O. Is the airway volume being correctly analyzed? *Am. J. Orthod. Dentofac. Orthop.* **2012**, *141*, 657–661. [CrossRef] [PubMed]
35. Van Eijnatten, M.; Koivisto, J.; Karhu, K.; Forouzanfar, T.; Wolff, J. The impact of manual threshold selection in medical additive manufacturing. *Int. J. Comput. Assist. Radiol. Surg.* **2017**, *12*, 607–615. [CrossRef] [PubMed]
36. Minnema, J.; van Eijnatten, M.; Kouw, W.; Diblen, F.; Mendrik, A.; Wolff, J. CT image segmentation of bone for medical additive manufacturing using a convolutional neural network. *Comput. Biol. Med.* **2018**, *103*, 130–139. [CrossRef]
37. ElShebiny, T.; Morcos, S.; El, H.; Palomo, J.M. Comparing different software packages for measuring the oropharynx and minimum cross-sectional area. *Am. J. Orthod. Dentofac. Orthop.* **2022**, *161*, 228–237.e32. [CrossRef]
38. Shi, H.; Scarfe, W.C.; Farman, A.G. Upper airway segmentation and dimensions estimation from cone-beam CT image datasets. *Int. J. Comput. Assist. Radiol. Surg.* **2006**, *1*, 177–186. [CrossRef]
39. Alsufyani, N.A.; Hess, A.; Noga, M.; Ray, N.; Al-Saleh, M.A.Q.; Lagrav'ere, M.O.; Major, P.W. New algorithm for semiautomatic segmentation of nasal cavity and pharyngeal airway in comparison with manual segmentation using cone-beam computed tomography. *Am. J. Orthod. Dentofac. Orthop.* **2016**, *150*, 703–712. [CrossRef]
40. Neelapu, B.C.; Kharbanda, O.P.; Sardana, V.; Gupta, A.; Vasamsetti, S.; Balachandran, R.; Rana, S.S.; Sardana, H.K. A pilot study for segmentation of pharyngeal and sino-nasal airway subregions by automatic contour initialization. *Int. J. Comput. Assist. Radiol. Surg.* **2017**, *12*, 1877–1893. [CrossRef]

Systematic Review

Surgical Treatment Options for Epiglottic Collapse in Adult Obstructive Sleep Apnoea: A Systematic Review

Kyriaki Vallianou [1] and Konstantinos Chaidas [2,*]

1   Ear, Nose, and Throat (ENT) Department, University Hospital of Larissa, 41334 Larissa, Greece
2   Ear, Nose, and Throat (ENT) Department, University Hospital of Alexandroupolis, Department of Medicine, Democritus University of Thrace, 68100 Alexandroupolis, Greece
*   Correspondence: konchaidas@gmail.com

Abstract: The critical role of epiglottis in airway narrowing contributing to obstructive sleep apnoea (OSA) and continuous positive airway pressure (CPAP) intolerance has recently been revealed. This systematic review was conducted to evaluate available surgical treatment options for epiglottic collapse in adult patients with OSA. The Pubmed and Scopus databases were searched for relevant articles up to and including March 2022 and sixteen studies were selected. Overall, six different surgical techniques were described, including partial epiglottectomy, epiglottis stiffening operation, glossoepiglottopexy, supraglottoplasty, transoral robotic surgery, maxillomandibular advancement and hypoglossal nerve stimulation. All surgical methods were reported to be safe and effective in managing selected OSA patients with airway narrowing at the level of epiglottis. The surgical management of epiglottic collapse can improve OSA severity or even cure OSA, but can also improve CPAP compliance. The selection of the appropriate surgical technique should be part of an individualised, patient-specific therapeutic approach. However, there are not enough data to make definitive conclusions and additional high-quality studies are required.

Keywords: epiglottis; epiglottic collapse; obstructive sleep apnoea; surgical treatment; surgery

Citation: Vallianou, K.; Chaidas, K. Surgical Treatment Options for Epiglottic Collapse in Adult Obstructive Sleep Apnoea: A Systematic Review. *Life* **2022**, *12*, 1845. https://doi.org/10.3390/life12111845

Academic Editor: Emanuela Locati

Received: 24 October 2022
Accepted: 9 November 2022
Published: 11 November 2022

Publisher's Note: MDPI stays neutral with regard to jurisdictional claims in published maps and institutional affiliations.

1. Introduction

Obstructive sleep apnoea (OSA) is a common sleep disorder characterised by repeated episodes of partial or complete airway collapse at various levels of the upper respiratory tract during sleep, which leads to a decrease or cessation of airflow [1]. OSA affects at least 7.8% of the adult population, with prevalence exceeding 50% in some countries [2], and is characterised by oxygen desaturation, sleep fragmentation and excessive daytime sleepiness [3]. Untreated OSA is an important risk factor of cardiovascular disease, arterial and pulmonary hypertension, arrythmias, diabetes and mortality [4].

The gold-standard method for OSA diagnosis is overnight polysomnography [3,5]. Airway obstruction can be single- or multi-level, including the velum, the oropharynx, the tongue base and/or the epiglottis. Epiglottic collapse was reported to concern a relatively small number of OSA patients and is often overlooked [6]. However, the actual prevalence seems to be higher, as the diagnosis of epiglottic collapse can elude doctors and be underestimated on awake endoscopy [7]. The addition of drug-induced sedation endoscopy (DISE) to the diagnostic approach helps surgeons in identifying the level of obstruction and provides the ability for more targeted treatment, increasing its success rate [8,9].

Continuous positive airway pressure (CPAP) is the first-line therapy in OSA, but is often associated with poor compliance [9]. Among alternative therapeutic modalities, surgical treatment constitutes an effective option in selected patients and especially in those with narrowing at the level of the epiglottis [10,11]. The aim of this systematic review is to provide an overview of the current literature regarding surgical treatment options in OSA patients with epiglottic collapse.

2. Materials and Methods

The Pubmed and Scopus databases were searched for relevant journal articles up to and including March 2022. Search terms included 'sleep apnoea', 'epiglottic collapse' or 'epiglottis' or 'epiglottic obstruction', 'treatment' or 'management' or 'therapy'. We aimed to identify all full-text articles that examined the surgical treatment options for treating epiglottic collapse in patients with obstructive sleep apnoea. The eligibility criterion for inclusion in the review was a specific focus on the surgical treatment of epiglottic collapse in an adult population (>16 years of age). Studies with no data on surgical management, articles concerning paediatric patients (≤16 years old), reviews and books were excluded. Non-English articles and animal and model studies were also excluded. Two authors independently performed the article search, article selection and data extraction. The reference lists of the chosen articles were manually searched to further identify relevant articles. The PRISMA guidelines were adapted for the current review.

3. Results

3.1. Search Results and Article Selection

The electronic database search identified 1181 articles. After the removal of 207 duplicates, the articles were screened by evaluating the titles and abstracts and selection was made based on the appliance of inclusion and exclusion criteria. A total of 169 articles were selected and full texts were retrieved. After further evaluation of their full text, 154 articles were excluded for the following reasons: language restrictions, full text unavailable, conservative treatment of epiglottic collapse, paediatric patients, animal studies, anatomic location of airway obstruction, books and reviews. In total, sixteen studies met the eligibility criteria, as one additional article was added from the reference search (Figure 1). The articles were categorised into three case reports (CRs), four case series (CSs), seven retrospective cohort studies (RCSs), and two prospective cohort studies (PCSs).

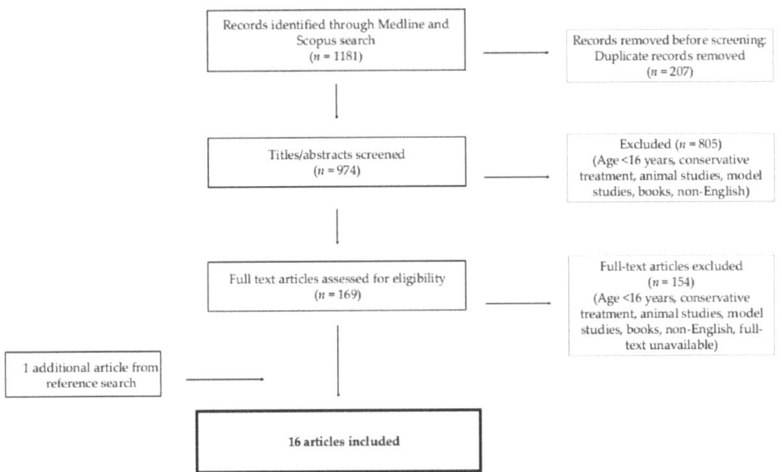

Figure 1. Literature search and article selection; *n*: number of studies.

The variations between the studies with regard to study type, patient characteristics and type of surgery are outlined in Table 1. The quality of each study was evaluated by using the GRADE (Grading of Recommendations Assessment, Development and Evaluation) system [12]. Additionally, Table 2 shows the preoperative and postoperative outcomes, including the Oxygen Desaturation Index (ODI), the Apnoea Hypopnoea Index (AHI) and the Epworth Sleepiness Scale (ESS) scores for each study, in order to evaluate the effectiveness of surgical treatment on epiglottic collapse.

Table 1. Individual study characteristics.

Study	Study Type	Patients Nr	Sex	Age (Years)	BMI (kg/m^2)	Physical Examination	Surgical Procedure	Follow-Up	Study Quality (GRADE)
Heiser [13]	CR	1	Male	64	25.9		Hypoglossal nerve stimulation	6 months	Very low
Verse [14]	CR	1	Male	70	24.2	Large epiglottis adhered to posterior pharyngeal wall	CO$_2$ partial epiglottic resection	7 days	Very low
Oluwasamni [15]	CS	4	Male	50–65		Floppy epiglottis	Endoscopic partial epiglottidectomy	2 months–3.5 years	Very low
Liu [16]	RCS	20	17 males/ 3 females	44 ± 12			Maxillomandibular advancement	6 months	Low
Li [17]	CR	1	Male	24		Long epiglottis touching the uvula and tilted posteriorly against the pharyngeal wall	Supraglottoplasty	6 months	Very low
Golz [18]	RCS	27	21 males/ 6 females	19–68	23.4 ± 4.2	Long, lax and flaccid epiglottis collapsing into the laryngeal inlet	CO$_2$ partial epiglottectomy	14–52 months (mean:32.3 months)	Low
Liu [19]	RCS	4/16	15 males/ 1 female	47 ± 10.9	29.4 ± 5.1	Partial collapse (anterioposterior:2, lateral:1), complete anterioposterior:1	Maxillomandibular advancement	6 months	Very low
Kayhan [20]	RCS		19 males/ 6 females	50.1± 8.5	30.7 ± 5.5		TORS	3 months	Low
Leone [21]	RCS	1/6	Male	58	25.3	Floppy epiglottis, epiglottis malacia	Epiglottis stiffening operation	8 months	Very low
Arora [22]	PCS	10/14	13 males/ 1 female	54.3 ± 14.6	28.7 ± 2.8	Concurrent epiglottic collapse	TORS	18.9 ± 6.2 months	Low
Jeong [23]	CS	2	Male	Pt1: 50 Pt2: 58	Pt1:29.1 Pt2:25	Complete epiglottic collapse	Partial epiglottectomy	1 year	Very low
Xiao [24]	RCS	13/48	32 males/ 16 females	57–69 (66)	28.6		Hypoglossal nerve stimulation	3 months	Low
Shehan [25]	CS	2	Male	Pt1: 60 Pt2: 55			Robotic-assisted epiglottopexy	Pt1: 1 year	Low
Salamanca [26]	RCS	14	13 males/ 1 female	47–76	25.6 (22.1–34)		Epiglottis stiffening operation	3 months	Low

Table 1. *Cont.*

Study	Study Type	Patients Nr	Sex	Age (Years)	BMI (kg/m^2)	Physical Examination	Surgical Procedure	Follow-Up	Study Quality (GRADE)
Catalfumo [27]	PCS	12		42.3 ± 14.6		Nine patients: stage 2 epiglottic position (45–90 degrees), Three patients: stage 3 epiglottic position (over 90 degrees)	CO$_2$ partial epiglottis resection	1 year	Low
Roustan [28]	CS	20	16 males/4 females	38–63			Transoral glossoepiglottopexy	6 months	Low

Values are given as number, mean ± SD or range; Nr: number, BMI: body mass index, GRADE: grading of recommendations assessment, development and evaluation, CR: case report, CS: case series, RCS: retrospective cohort study, PCS: prospective cohort study, CO$_2$: carbon dioxide laser, TORS: transoral robotic surgery, Pt: patient.

Table 2. Preoperative and Postoperative Outcomes.

Study	Preoperative ODI (Events/h)	Preoperative AHI (Episodes/h)	Preoperative ESS	Preoperative SaO$_2$ (%)	Postoperative ODI (Events/h)	Postoperative AHI (Episodes/h)	Postoperative ESS
Heiser [13]		36.3		Min: 76		20.7	
Verse [14]	11.1	4.8	7	Min: 77	3.8	0.4	
Oluwasamni [15]							
Liu [16]	38.7 ± 30.3	53.6 ± 26.6			8.1 ± 9.2	9.5 ± 7.4	
Li [17]	21		9	Min: 89.1	6.9		5
Golz [18]	26–65 (45 ± 14.6)			Min: 58–87 (Mean: 66 ± 17.6)	14 ± 5.1		
Liu [19]	45 ± 29.7	59.8 ± 25.6	19.5 ± 2.9	Min: 80.8 ± 7.6	5.7 ± 4.4	9.3 ± 7.1	7.1 ± 2.6
Kayhan [20]		28.7 ± 17.8	13.5 ± 2.8	Min: 80.7 ± 7.6		9.4 ± 12.4	3.4 ± 1.6
Leone [21]		47.7				4.7	
Arora [22]		35.6 ± 19.7	14.9 ± 5	Mean: 92.9 ± 1.8		21.2 ± 24.6	All patients: normal (<10)
Jeong [23]	Pt1: 46.1	Pt1: 57.8, Pt2: 28.4		Pt1 min: 73, Pt2 min: 69		Pt1: 50.5, Pt2: 25.7	
Xiao [24]							
Shehan [25]		Pt1: 35, Pt2: 28	Pt2: 8			Pt1: 31, Pt2: 6	Pt1: 3, Pt2: 2
Salamanca [26]		0.1–57.6	1–9			0.2–6.8	1–7
Catalfumo [27]		42 ± 16.4		Min: 68 ± 8.6	12 ± 4.6	8 ± 3.2	
Roustan [28]	23 ± 14.3	23.6 ± 6.5	16.5 ± 4.3	Mean: 86.9 ± 2.3		5.2 ± 3.2	3.1 ± 2.5

Values are given as mean ± SD or range; ODI: oxygen desaturation index, h: hour, AHI: apnoea hypopnea index, ESS: Epworth Sleepiness Scale, SaO$_2$: oxygen saturation, Min: minimum, Pt: patient.

3.2. Partial Epiglottectomy

A total of three studies were identified that reported CO_2 partial epiglottis resection for the treatment of epiglottic collapse in OSA patients [14,18,27]. In a study published by Catalfumo et al., twelve patients with epiglottis collapse underwent partial epiglottectomy, with no serious complications. The results indicate that this operation can increase the OSA treatment success rate by 10–15%, with a significant improvement in oxygen blood saturation, apnoea duration and the apnoea hypopnea index (AHI) score after surgery [27]. Golz et al. performed U-shaped partial epiglottectomy with a CO_2 laser in 27 adults with OSA and laryngomalacia, uneventfully. Postoperative sleep studies demonstrated a statistically significant improvement in 85% of patients and complete relief of their respiratory symptoms. A significant decrease in the respiratory disturbance index (RDI) was achieved in 77.8% of the patients [18]. Verse et al. [14] published a case report of a 70-year-old man who underwent a partial epiglottectomy using the CO_2 laser, due to a very large epiglottis. After the surgery, the patient regained the ability to sleep in the supine position with a reduced AHI score and complete disappearance of all respiratory disturbances [14]. Not only the CO_2 laser, but also monopolar diathermy can be used for partial epiglottectomy [15,23]. Oluwasamni et al. reported four cases undergoing endoscopic partial epiglottectomy using monopolar diathermy to treat a floppy epiglottis. This was found to be safe and effective with minimal morbidity [15]. Similarly, in a study published by Jeong et al., the authors used monopolar electrocautery to proceed to partial resection of the epiglottis, in order to improve CPAP usage. The patients were relieved from the feeling of suffocation with a reduction in the AHI score and satisfactory use of CPAP [23].

3.3. Epiglottis Stiffening Operation

The literature search revealed two articles that described the epiglottis stiffening operation (ESO) as a therapeutic approach to epiglottic collapse in OSA patients [21,26]. Salamanca et al. [26] used suction cautery to cauterise the lower half of the lingual surface of the epiglottis in the area between the lateral glosso-epiglottic folds to induce stiffening and scar retraction of tissues as a result of secondary healing. The authors highlighted the importance of reaching the perichondrium of the lingual side of the epiglottis, to induce stiffening in the direction of median thyroepiglottic ligament. ESO was performed in 14 patients without complications such as dysphagia or aspiration [26]. Leone et al. published a retrospective study, in which one patient developed a sleep-related breathing disorder, due to a floppy epiglottis as a consequence of chemo- and radiotherapy for head and neck cancer. The patient underwent an epiglottis stiffening operation without postoperative complications and reached a resolution of OSA with normalisation of the AHI after surgery (from 47.7 episodes/h prior to surgery, the AHI decreased to 4.7 episodes/h, postoperatively) [21].

3.4. Glossoepiglottopexy

Roustan et al. developed a surgical technique that provides a support to the epiglottis without destroying its function during swallowing. They analysed a group of 20 patients who underwent glossoepiglottopexy using a CO_2 laser and pharyngoplasty. There was a significant reduction in the ESS score, AHI and ODI with normal swallowing in all patients, postoperatively [28].

3.5. Supraglottoplasty

Supraglottoplasty with the intraoral and laryngoscopic approach was carried out by Li et al. in one patient with OSA caused by laryngomalacia [17]. There was a good response to treatment with improvement in snoring, daytime sleepiness and sleep apnoea. Postoperative endoscopy revealed a reduction in the size of the epiglottis and arytenoids without collapse of the supraglottic tissue.

3.6. Transoral Robotic Surgery (TORS)

Three studies published the results of robotic-assisted surgery in the epiglottis for OSA [20,22,25]. Shehan et al. [25] reported the first robotic-assisted epiglottopexy in the adult otolaryngology literature. Namely, the authors described two cases with epiglottic collapse undergoing robotic-assisted epiglottopexy. The first patient showed a decrease in the AHI and ODI and the second patient had a significant decrease in the AHI and ESS [25]. Kayhan et al. [20] evaluated the results of combined multilevel surgery with transoral robotic surgery (TORS) in patients with OSA and multilevel airway obstruction. DISE was performed in all patients in order to identify the level of obstruction and determine the type of surgery required. A total of 24 patients underwent a base of tongue (BOT) reduction and epiglottoplasty. There was no need for tracheostomy in any of the patients. Sleep apnoea was cured in 72% of the patients, whereas an additional 8% of the patients met the criteria for surgical success. Another prospective study published by Arora et al. [22] included 10 patients who underwent TORS for tongue base reduction and epiglottoplasty. A 64% success rate was achieved with a normal postoperative polysomnography in 36% of cases at six months. There was a 51% reduction in the mean AHI score.

3.7. Maxillomandibular Advancement

Our search identified two studies by Liu et al. in which patients with OSA and epiglottic collapse underwent maxillomandibular advancement (MMA) [16,19]. The first retrospective cohort study included four patients with partial or complete epiglottic collapse and a mean preoperative AHI score of 59.8 episodes per hour. Post-MMA, the AHI and ODI scores had a statistically significant reduction, whereas postoperative sleep endoscopy showed an improvement in the collapsibility of the epiglottis [19]. Subsequently, Liu et al. published a retrospective cohort study of 20 OSA patients undergoing MMA. A total of 6 out of 20 patients were found to have narrowing at the level of the epiglottis at the preoperative assessment, which was resolved in half (three) of them after surgery. In contrast, three patients had a residual epiglottic collapse despite surgery. The authors found that MMA increases the stability of the lateral pharyngeal wall, followed by the velum and the tongue base. The results were assessed with DISE and computational fluid dynamics [16].

3.8. Hypoglossal Nerve Stimulation

A retrospective cohort study by Xiao et al. included 13 patients who underwent hypoglossal nerve stimulation (HNS) as a treatment for partial or complete epiglottic collapse and obstructive sleep apnoea. The authors reported a significant improvement in the AHI, ODI and ESS three months after HNS implantation [24]. In a study by Heiser et al., a 64-year-old male presented with residual OSA with an increased AHI and evidence of a floppy epiglottis six months after upper airway stimulation [13]. A change of the electrode configuration for stimulation from bipolar to monopolar produced a clear opening of the epiglottis and the patient improved significantly.

4. Discussion

According to the guidelines published by the American Academy of Sleep Medicine, a referral to a sleep surgeon should be considered for patients with OSA and a BMI of less than 40 kg/m^2 who are intolerant or unaccepting of CPAP [29]. In adult OSA, airway obstruction is often present at multiple levels and a thorough assessment of the upper respiratory tract is necessary. DISE provides valuable information about the presence, level and type of obstruction and can guide the selection of the appropriate surgical method, increasing its success rate [9]. The role of DISE is crucial, especially in the presence of epiglottic collapse, which is usually difficult to identify on awake endoscopy. Although the role of the epiglottis in OSA was underestimated for many years, several recent reports confirm the increased prevalence of this type of obstruction and highlight the importance of targeted management in order to resolve OSA or improve CPAP compliance [7]. Our review demonstrates that various surgical methods with satisfactory results and minimum morbidity exist. The selection of the optimal surgical option should primarily be guided by the type of underlying pathology, although other factors, such as patient characteristics and preference, and equipment availability, should also be considered.

The type of epiglottic collapse can be divided into anterioposterior, which is the most common, and lateral collapse [30]. Before the advent of DISE, epiglottic collapse evaluated by an awake clinical examination was reported to occur in 11.4% of OSA patients [28], but its prevalence has increased since the wider use of DISE [30]. A study by Fernandez-Julian et al. found that the epiglottis was involved in the obstruction of 24.1–28.4% of the patients according to an awake examination, a rate that increased to 36.4% based on DISE [31]. Lan et al. [32] noted anterioposterior or lateral epiglottic collapse in 42.2% of their patients. Another study using DISE in 324 patients reported a floppy epiglottis in 18.5% of them [33]. In most cases, epiglottic collapse co-exists with obstruction at other sites, whereas isolated epiglottic collapse is seen in a significantly smaller number of patients, with a rate ranging between 3.5% and 14.4% [18,34,35]. The prevalence of epiglottic collapse is significantly higher when evaluating OSA patients who have previously failed upper airway surgery, ranging from 44–72.9% [36,37].

The pathophysiology of epiglottic collapse remains not well understood and several mechanisms have been proposed to explain this phenomenon. Epiglottic collapse may occur secondary to anterioposterior collapse of the base of the tongue, pushing the epiglottis backwards, or due to underdevelopment of the epiglottis, which leads to lateral collapse. An underdeveloped epiglottis can cause laryngomalacia and sleep apnoea. Another described mechanism is the complete isolated anterioposterior epiglottic collapse occurring during inspiration, also known as the 'trapdoor phenomenon' [33]. In this case, the epiglottis prolapses into the posterior pharyngeal wall during inspiration, causing airway obstruction. Moreover, a previous history of radiotherapy for oropharyngeal or laryngeal cancer has been reported to increase the risk for OSA by causing oedema and malacia of the epiglottis and/or a floppy epiglottis [38].

The relationship between the shape of the epiglottis and its potential collapse is still controversial. Sung et al. assessed 11 cases with isolated epiglottic collapse and 44 controls and found no differences in terms of epiglottic shape or curvature between the two groups [35]. Another study showed no significant correlation between the position of the epiglottis and the presence of collapse [33]. In contrast, Kanemaru et al. identified a positive association between a concave posterior surface of the epiglottis and the degree of airway collapse and, thus, the severity of OSA [39]. Catalfumo et al. [27] compared the results of full polysomnography studies and found a correlation between the severity of OSA and the position of the epiglottis. Kuo et al. [40] pointed out the connection between the epiglottic length and epiglottic collapse, which can be screened with a CT scan or cephalometry. An epiglottic length of more than 1.66 cm was indicative of collapse of the epiglottis. Additionally, high hyoid mobility was correlated with epiglottic collapse in OSA patients. In contrast, the epiglottic angle did not seem to play a role [40].

Satisfactory management in OSA patients presupposes the identification of the location of the upper airway obstruction. Several diagnostic methods have been used to assess upper airway patency and identify associated pathology in patients with OSA, including awake flexible endoscopy, computed tomography (CT), magnetic resonance imaging (MRI), cephalometry, intrapharyngeal pressure manometry and DISE. Since its first introduction in 1991, DISE has gained popularity, as it allows a direct dynamic evaluation of the upper airway during drug-induced "sleep" [6]. This technique is a useful method of assessing the location, severity and pattern of airway obstructions. To better describe DISE findings in a standardised way, the VOTE (velum, oropharynx, tongue base, epiglottis) classification has been widely used with very good intra- and inter-rater reliability [8]. All patients should undergo an awake fibre-optic nasolaryngoscopy prior to any surgical procedure to enable the surgeon to fully assess the upper airway and identify any additional pathology. Those with a high suspicion of epiglottic collapse should also undergo DISE, as it is challenging to fully assess this type of obstruction on awake endoscopy [37].

Although CPAP is the first-line therapy in OSA, its efficacy seems to be limited in patients with epiglottic collapse [35]. The positive airway pressure can push the epiglottis downward into the laryngeal inlet, leading to significant narrowing of the upper airway, worsening of OSA and/or CPAP intolerance [30]. A recent study by Sung et al. suggests that patients with epiglottic collapse have a higher CPAP adherence failure rate than patients without epiglottic collapse [41]. Salamanca et al. and Shehan et al. also agree that a poor response to CPAP in patients with epiglottic collapse occurs because the positive pressure may worsen the epiglottic obstruction [25,26]. A collapsing epiglottis has been found in 15–31.4% of adult patients with OSA who did not tolerate CPAP therapy or in whom CPAP treatment was ineffective [28,42–44]. Kim et al. confirmed this assertion by publishing an article with an OSA patient with worsening epiglottic collapse during CPAP application in a video presentation [45]. In case of residual obstruction after CPAP, an upper airway reassessment should be performed to determine the presence and type of epiglottic collapse [46], as these patients require different management.

Positional therapy and mandibular advancement devices are among the non-surgical treatment options that have been suggested to manage epiglottic collapse. Both treatment modalities seem to improve airway patency in those patients but only in mild cases, as in the presence of severe OSA, additional treatment is usually required [45,47].

In contrast to conservative management, surgical therapy of epiglottic collapse seems to be more effective. Several surgical techniques aiming to resolve epiglottic narrowing have been reported in the literature with overall good results. The available operations include less invasive techniques, such as partial epiglottectomy, glossoepiglottopexy and supraglottoplasty, and relatively more aggressive techniques, including transoral robotic surgery, maxillomandibular advancement and hypoglossal nerve stimulation. However, the use of new technologies, such as diathermy and CO_2 and thulium lasers, along with the gradual increase in surgical experience, have improved the safety of the operations [30].

It is known that the epiglottis participates in swallowing and is also involved in preventing food aspiration by closing the laryngeal aditus during swallowing. Although the main objective of surgical management is the improvement of upper airway patency by resolving epiglottic collapse, the function of the epiglottis should also be preserved.

Partial epiglottectomy is an effective tool for patients with laryngomalacia or trapdoor epiglottis. A carbon dioxide laser and monopolar diathermy have both been cited to be useful in the surgical treatment of OSA caused by laryngomalacia [15,27]. Partial epiglottectomy with a CO_2 laser has also been used to excise the redundant mucosa of the arytenoids. This instrument provides a high degree of precision, while maintaining homeostasis and minimising postoperative oedema. The technique was found to be safe without complications. By cutting out the upper-middle one third to half of the epiglottis, the aerodynamic shape of the hypopharynx is changed, resolving the obstruction caused by the epiglottis.

It is often challenging to determine the optimal volume of effective epiglottic resection without postoperative complications, as excessive epiglottic resection can cause aspiration, whereas insufficient resection carries the risk of residual obstruction. Some studies suggest leaving a residual 3–4 mm rim of healthy mucosa along the entire profile of the epiglottis [14,27,40]. Bartolomeo et al. [48] report that V-shaped partial epiglottectomy minimises the risk of aspiration, while ensuring satisfactory airflow through the epiglottic V during epiglottic movement.

The epiglottis stiffening operation (ESO) was first described by Salamanca et al. [26] in 2019. Following this technique, stiffening and scar retraction leads to flexion of the epiglottis towards the tongue base by secondary intention. The authors suggested leaving some healthy tissue along the free border of the epiglottis to allow the activation of reflexes. Overall, the ESO was found to be a safe and effective procedure with a shorter healing time than partial epiglottectomy [26].

Another minimally invasive technique which can achieve resolution of epiglottic collapse, while preserving the function of the epiglottis, is glossoepiglottopexy. Transoral glossoepiglottopexy constitutes a safe and effective surgical treatment option for adults with OSA and epiglottic collapse and is associated with a lower risk of complications compared to partial epiglottectomy [27,28]. This method provides stable support to the epiglottis, protects its function during swallowing and creates a barrier for posterior falling off the tongue base by reinforcing the wall of the airway [28].

Transoral robotic surgery (TORS) was first reported in 2010 as a modification of open tongue base reduction and hyoid epiglottopexy to treat OSA [49]. It has been demonstrated that TORS of the tongue base with or without epiglottoplasty by using a CO_2 or thulium laser is a safe and effective alternative treatment option for selected patients, when other treatment options failed. The ideal candidates are patients with an obstruction at the level of tongue base and/or epiglottis and the procedure can be completed without major complications or the need for tracheostomy or open surgery [22]. Although the surgeon has no haptic feedback intra-operatively, TORS provides an excellent approach at the hypopharynx with three-dimensional visualisation of the surgical field [20].

Maxillomandibular advancement constitutes an alternative surgical treatment for patients with an intolerance to CPAP therapy, particularly for patients with severe lateral pharyngeal and epiglottic collapse diagnosed on DISE. MMA is considered one of the most effective treatment options for patients with OSA, but is associated with significantly higher complication rates compared to other surgical options [50].

Non-obese patients with a history of previous CPAP failure or intolerance should also be screened for upper airway stimulation [13,24]. Stimulation of the hypoglossal nerve and activation of the genioglossus muscle unbars the base of the tongue and the soft palate. Additional stimulation settings, such as changing the configuration of the stimulation electrode, may optimise muscle recruitment to favour upper airway dilation at the levels of the soft palate, tongue and epiglottis [13]. Despite the relatively excessive surgical dissection required for implantation, hypoglossal nerve stimulation is a safe and effective method with promising outcomes and low associated morbidity [24].

The aim of this systematic review was to assess the efficacy of surgical therapy in patients with OSA and epiglottic collapse. The current evidence shows that several surgical options with satisfactory outcomes and safety profiles exist. However, this review carries out certain limitations and conclusions should be made with caution. First, a systematic review was conducted but not a meta-analysis. The studies included were all either retrospective or prospective observational studies. None of the studies was randomised or multi-centre. Additionally, the outcomes were based on small sample sizes. There was also a variation in the methodology and the duration of follow-up assessment after surgery. This literature review included studies specifically mentioning the effect of surgical treatment in patients with OSA and epiglottic collapse. However, it should be taken into consideration that other techniques such as tongue base advancement may also affect airway patency at the level of the epiglottis. Our literature search revealed that most studies have focused on the impact of a surgical technique on the level of the tongue base or on upper airway patency in general, without specifically evaluating and reporting outcomes on epiglottic collapse. These studies are beyond the scope of this review and were excluded.

5. Conclusions

Since DISE has become a popular method for upper airway examination, the critical role of the epiglottis in airway narrowing contributing to OSA has been revealed. Unfortunately, CPAP intolerance or failure is relatively common in patients with epiglottic collapse and, thus, alternative treatment options should be considered. Several surgical techniques have been described in the literature, with overall satisfactory results and safety profiles. The surgical management of epiglottic collapse can improve OSA severity or even cure OSA but can also improve CPAP compliance. Moreover, as upper airway obstruction is often multilevel, epiglottic surgery can also be combined with other upper airway procedures. The selection of the appropriate surgical technique should be made based on the type of airway obstruction, patient characteristics and preferences, surgical skills and equipment availability, as part of an individualised, patient-specific therapeutic approach.

Nevertheless, our findings should be evaluated with caution due to the low quality of the included studies. For that reason, there is a need for high-quality randomised trials with large sample sizes to allow us to safely determine the effect of surgery in patients with OSA and epiglottic collapse.

Author Contributions: Conceptualization, K.C.; methodology, K.C.; software, K.V., K.C.; validation, K.V., K.C.; formal analysis, K.V., K.C.; investigation, K.V., K.C.; resources, K.V., K.C.; data curation: K.V., K.C.; writing-original draft preparation, K.V., K.C.; writing-review and editing, K.C.; visualization, K.V., K.C.; supervision, K.C.; project administration: K.C. All authors have read and agreed to the published version of the manuscript.

Funding: This research received no external funding.

Informed Consent Statement: Not applicable.

Data Availability Statement: Not applicable.

Conflicts of Interest: The authors declare no conflict of interest.

References

1. Faber, J.; Faber, C.; Faber, A.P. Obstructive sleep apnea in adults. *Dent. Press J. Orthod.* **2019**, *24*, 99–109. [CrossRef] [PubMed]
2. Benjafield, A.V.; Ayas, N.T.; Eastwood, P.R.; Heinzer, R.; Ip, M.S.M.; Morrell, M.J.; Nunez, C.M.; Patel, S.R.; Penzel, T.; Pépin, J.-L.; et al. Estimation of the global prevalence and burden of obstructive sleep apnoea: A literature-based analysis. *Lancet Respir. Med.* **2019**, *7*, 687–698. [CrossRef]
3. American Academy of Sleep Medicine. *The International Classification of Sleep Disorders, Revised: Diagnostic and Coding Manual*; American Academy of Sleep Medicine: Chicago, IL, USA, 2001.
4. Shahar, E.; Whitney, C.W.; Redline, S.; Lee, E.T.; Newman, A.B.; Nieto, F.J.; O'Connor, G.T.; Boland, L.L.; Schwartz, J.E.; Samet, J.M. Sleep-disordered breathing and cardiovascular disease: Cross-sectional results of the Sleep Heart Health Study. *Am. J. Respir. Crit. Care Med.* **2001**, *163*, 19–25. [CrossRef]
5. Sateia, M. International classification of sleep disorders-third edition: Highlights and modifications. *Chest* **2014**, *146*, 1387–1394. [CrossRef]
6. Amos, J.M.; Durr, M.L.; Nardone, H.C.; Baldassari, C.M.; Duggins, A.; Ishman, S.L. Systematic Review of Drug-Induced Sleep Endoscopy Scoring Systems. *Otolaryngol. Neck Surg.* **2017**, *158*, 240–248. [CrossRef] [PubMed]
7. Sung, C.M.; Tan, S.N.; Shin, M.-H.; Lee, J.; Kim, H.C.; Lim, S.C.; Yang, H.C. The Site of Airway Collapse in Sleep Apnea, Its Associations with Disease Severity and Obesity, and Implications for Mechanical Interventions. *Am. J. Respir. Crit. Care Med.* **2021**, *204*, 103–106. [CrossRef] [PubMed]
8. Kezirian, E.J.; Hohenhorst, W.; de Vries, N. Drug-induced sleep endoscopy: The VOTE classification. *Eur. Arch. Otorhinolaryngol.* **2011**, *268*, 1233–1236. [CrossRef]
9. Stuck, B.A.; Maurer, J.T. Airway evaluation in obstructive sleep apnea. *Sleep Med. Rev.* **2008**, *12*, 411–436. [CrossRef]
10. Sin, D.D.; Mayers, I.; Man, G.C.; Pawluk, L. Long-term compliance rates to continuous positive airway pressure in obstructive sleep apnea: A population-based study. *Chest* **2002**, *121*, 430–435. [CrossRef]
11. Certal, V.; Nishino, N.; Camacho, M.; Capasso, R. Reviewing the Systematic Reviews in OSA Surgery. *Otolaryngol. Neck Surg.* **2013**, *149*, 817–829. [CrossRef]
12. Guyatt, G.H.; Oxman, A.D.; Vist, G.E.; Kunz, R.; Falck-Ytter, Y.; Alonso-Coello, P.; Schünemann, H.J.; GRADE Working Group. GRADE: An emerging consensus on rating quality of evidence and strength of recommendations. *BMJ* **2008**, *336*, 924–926. [CrossRef] [PubMed]
13. Heiser, C. Advanced titration to treat a floppy epiglottis in selective upper airway stimulation. *Laryngoscope* **2016**, *126* (Suppl. S7), S22–S24. [CrossRef] [PubMed]
14. Verse, T.; Pirsig, W. Age-related changes in the epiglottis causing failure of nasal continuous positive airway pressure therapy. *J. Laryngol. Otol.* **1999**, *113*, 1022–1025. [CrossRef] [PubMed]
15. Oluwasanmi, A.F.; Mal, R.K. Diathermy epiglottectomy: Endoscopic technique. *J. Laryngol. Otol.* **2001**, *115*, 289–292. [CrossRef]
16. Liu, S.Y.-C.; Huon, L.-K.; Iwasaki, T.; Yoon, A.; Riley, R.; Powell, N.; Torre, C.; Capasso, R. Efficacy of Maxillomandibular Advancement Examined with Drug-Induced Sleep Endoscopy and Computational Fluid Dynamics Airflow Modeling. *Otolaryngol. Neck Surg.* **2016**, *154*, 189–195. [CrossRef]
17. Li, H.-Y.; Fang, T.-J.; Lin, J.-L.; Lee, Z.-L.; Lee, L.-A. Laryngomalacia causing sleep apnea in an osteogenesis imperfecta patient. *Am. J. Otolaryngol.* **2002**, *23*, 378–381. [CrossRef]
18. Golz, A.; Goldenberg, D.; Westerman, S.T.; Catalfumo, F.J.; Netzer, A.; Westerman, L.M.; Joachims, H.Z. Laser partial epiglottidectomy as a treatment for obstructive sleep apnea and laryngomalacia. *Ann. Otol. Rhinol. Laryngol.* **2000**, *109 Pt 1*, 1140–1145. [CrossRef]

19. Liu, S.Y.; Huon, L.K.; Powell, N.B.; Riley, R.; Cho, H.G.; Torre, C.; Capasso, R. Lateral Pharyngeal Wall Tension After Maxillo-mandibular Advancement for Obstructive Sleep Apnea Is a Marker for Surgical Success: Observations from Drug-Induced Sleep Endoscopy. *J. Oral Maxillofac. Surg.* **2015**, *73*, 1575–1582. [CrossRef]

20. Kayhan, F.T.; Kaya, K.H.; Koç, A.K.; Yegin, Y.; Yazici, Z.M.; Türkeli, S.; Sayin, I. Multilevel Combined Surgery with Transoral Robotic Surgery for Obstructive Sleep Apnea Syndrome. *J. Craniofac. Surg.* **2016**, *27*, 1044–1048. [CrossRef]

21. Leone, F.; Marciante, G.A.; Re, C.; Bianchi, A.; Costantini, F.; Salamanca, F.; Salvatori, P. Obstructive sleep apnoea after radiotherapy for head and neck cancer. *Acta Otorhinolaryngol. Ital.* **2020**, *40*, 338–342. [CrossRef]

22. Arora, A.; Chaidas, K.; Garas, G.; Amlani, A.; Darzi, A.; Kotecha, B.; Tolley, N.S. Outcome of TORS to tongue base and epiglottis in patients with OSA intolerant of conventional treatment. *Sleep Breath.* **2016**, *20*, 739–747. [CrossRef] [PubMed]

23. Jeong, S.H.; Man Sung, C.; Lim, S.C.; Yang, H.C. Partial epiglottectomy improves residual apnea-hypopnea index in patients with epiglottis collapse. *J. Clin. Sleep Med.* **2020**, *16*, 1607–1610. [CrossRef] [PubMed]

24. Xiao, R.; Trask, D.K.; Kominsky, A.H. Preoperative Predictors of Response to Hypoglossal Nerve Stimulation for Obstructive Sleep Apnea. *Otolaryngol. Head Neck Surg.* **2020**, *162*, 400–407. [CrossRef] [PubMed]

25. Shehan, J.N.; Du, E.; Cohen, M.B. Robot-assisted epiglottopexy as a method for managing adult obstructive sleep apnea. *Am. J. Otolaryngol.* **2020**, *41*, 102742. [CrossRef]

26. Salamanca, F.; Leone, F.; Bianchi, A.; Bellotto, R.G.S.; Costantini, F.; Salvatori, P. Surgical treatment of epiglottis collapse in obstructive sleep apnoea syndrome: Epiglottis stiffening operation. *Acta Otorhinolaryngol. Ital.* **2019**, *39*, 404–408. [CrossRef]

27. Catalfumo, F.; Golz, A.; Westerman, S.; Gilbert, L.; Joachims, H.; Goldenberg, D. The epiglottis and obstructive sleep apnoea syndrome. *J. Laryngol. Otol.* **1998**, *112*, 940–943. [CrossRef]

28. Roustan, V.; Barbieri, M.; Incandela, F.; Missale, F.; Camera, H.; Braido, F.; Mora, R.; Peretti, G. Transoral glossoepiglottopexy in the treatment of adult obstructive sleep apnoea: A surgical approach. *Acta Otorhinolaryngol. Ital.* **2018**, *38*, 38–44. [CrossRef]

29. Kent, D.; Stanley, J.; Aurora, R.N.; Levine, C.; Gottlieb, D.J.; Spann, M.D.; Torre, C.A.; Green, K.; Harrod, C.G. Referral of adults with obstructive sleep apnea for surgical consultation: An American Academy of Sleep Medicine clinical practice guideline. *J. Clin. Sleep Med.* **2021**, *17*, 2499–2505. [CrossRef]

30. Torre, C.; Camacho, M.; Liu, S.Y.-C.; Huon, L.-K.; Capasso, R. Epiglottis collapse in adult obstructive sleep apnea: A systematic review. *Laryngoscope* **2016**, *126*, 515–523. [CrossRef]

31. Fernández-Julián, E.; García-Pérez, M.A.; García-Callejo, J.; Ferrer, F.; Martí, F.; Marco, J. Surgical planning after sleep versus awake techniques in patients with obstructive sleep apnea. *Laryngoscope* **2014**, *124*, 1970–1974. [CrossRef]

32. Lan, M.C.; Liu, S.Y.; Lan, M.Y.; Modi, R.; Capasso, R. Lateral pharyngeal wall collapse associated with hypoxemia in obstructive sleep apnea. *Laryngoscope* **2015**, *125*, 2408–2412. [CrossRef] [PubMed]

33. Vonk, P.E.; Ravesloot, M.J.L.; Kasius, K.M.; van Maanen, J.P.; de Vries, N. Floppy epiglottis during drug-induced sleep endoscopy: An almost complete resolution by adopting the lateral posture. *Sleep Breath.* **2020**, *24*, 103–109. [CrossRef] [PubMed]

34. Woodson, B.T. A method to describe the pharyngeal airway. *Laryngoscope* **2015**, *125*, 1233–1238. [CrossRef] [PubMed]

35. Sung, C.M.; Kim, H.C.; Yang, H.C. The clinical characteristics of patients with an isolate epiglottic collapse. *Auris Nasus Larynx* **2020**, *47*, 450–457. [CrossRef] [PubMed]

36. Dieleman, E.; Veugen, C.C.A.F.M.; Hardeman, J.A.; Copper, M.P. Drug-induced sleep endoscopy while administering CPAP therapy in patients with CPAP failure. *Sleep Breath.* **2021**, *25*, 391–398. [CrossRef]

37. Kim, D.K.; Lee, J.W.; Lee, J.H.; Lee, J.S.; Na, Y.S.; Kim, M.J.; Lee, M.J.; Park, C.H. Drug induced sleep endoscopy for poor-responder to uvulopalatopharyngoplasty in patient with obstructive sleep apnea patients. *Korean J. Otorhinolaryngol.-Head Neck Surg.* **2014**, *57*, 96–102. [CrossRef]

38. Huyett, P.; Kim, S.; Johnson, J.T.; Soose, R.J. Obstructive sleep apnea in the irradiated head and neck cancer patient. *Laryngoscope* **2017**, *127*, 2673–2677. [CrossRef]

39. Kanemaru, S.-I.; Kojima, H.; Fukushima, H.; Tamaki, H.; Tamura, Y.; Yamashita, M.; Umeda, H.; Ito, J. A case of floppy epiglottis in adult: A simple surgical remedy. *Auris Nasus Larynx* **2007**, *34*, 409–411. [CrossRef]

40. Kuo, I.C.; Hsin, L.J.; Lee, L.A.; Fang, T.J.; Tsai, M.S.; Lee, Y.C.; Shen, S.C.; Li, H.Y. Prediction of Epiglottic Collapse in Obstructive Sleep Apnea Patients: Epiglottic Length. *Nat. Sci. Sleep.* **2021**, *13*, 1985–1992. [CrossRef]

41. Sung, C.M.; Kim, H.C.; Kim, J.; Kim, J.G.; Lim, S.C.; Shin, M.H.; Nam, K.; Lee, J.; Vena, D.; White, D.P.; et al. Patients with Epiglottic Collapse are Less Adherent to Autotitrating Positive Airway Pressure Therapy for Obstructive Sleep Apnea. *Ann. Am. Thorac. Soc.* **2022**, *19*, 1907–1912. [CrossRef]

42. Dedhia, R.C.; Rosen, C.A.; Soose, R.J. What is the role of the larynx in adult obstructive sleep apnea? *Laryngoscope* **2014**, *124*, 1029–1034. [CrossRef] [PubMed]

43. Kwon, E.O.; Jung, S.Y.; Al-Dilaijan, K.; Min, J.Y.; Lee, K.H.; Kim, S.W. Is Epiglottis Surgery Necessary for Obstructive Sleep Apnea Patients with Epiglottis Obstruction? *Laryngoscope* **2019**, *129*, 2658–2662. [CrossRef] [PubMed]

44. Waters, T. Alternative interventions for obstructive sleep apnea. *Clevel. Clin. J. Med.* **2019**, *86*, 34–41. [CrossRef]

45. Kim, H.Y.; Sung, C.M.; Jang, H.B.; Kim, H.C.; Lim, S.C.; Yang, H.C. Patients with epiglottic collapse showed less severe obstructive sleep apnea and good response to treatment other than continuous positive airway pressure: A case-control study of 224 patients. *J. Clin. Sleep Med.* **2021**, *17*, 413–419. [CrossRef] [PubMed]

46. Bae, M.R.; Chung, Y.-S. Epiglottic Collapse in Obstructive Sleep Apnea. *Sleep Med. Res.* **2021**, *12*, 15–19. [CrossRef]

7. Kent, D.T.; Rogers, R.; Soose, R.J. Drug-induced sedation endoscopy in the evaluation of OSA patients with incomplete oral appliance therapy response. *Otolaryngol. Head Neck Surg.* **2015**, *153*, 302–307. [CrossRef]
8. Bartolomeo, M.; Bigi, A.; Pelliccia, P.; Makeieff, M. Surgical treatment of a case of adult epiglottic laryngomalacia. *Eur. Ann. Otorhinolaryngol. Head Neck Dis.* **2015**, *132*, 45–47. [CrossRef]
9. Vicini, C.; Dallan, I.; Canzi, P.; Frassineti, S.; La Pietra, M.G.; Montevecchi, F. Transoral robotic tongue base resection in obstructive sleep apnoea-hypopnoea syndrome: A preliminary report. *ORL J. Otorhinolaryngol. Relat. Spec.* **2010**, *72*, 22–27. [CrossRef]
10. Zhou, N.; Ho, J.-P.; Huang, Z.; Spijker, R.; de Vries, N.; Aarab, G.; Lobbezoo, F.; Ravesloot, M.; de Lange, J. Maxillomandibular advancement versus multilevel surgery for treatment of obstructive sleep apnea: A systematic review and meta-analysis. *Sleep Med. Rev.* **2021**, *57*, 101471. [CrossRef]

MDPI AG
Grosspeteranlage 5
4052 Basel
Switzerland
Tel.: +41 61 683 77 34

Life Editorial Office
E-mail: life@mdpi.com
www.mdpi.com/journal/life